The Marine Algae of Virginia

The Marine Algae of Virginia

Harold J. Humm

Special Papers in Marine Science, Number 3

Published for the

Virginia Institute of Marine Science
Gloucester Point

by the University Press of Virginia, Charlottesville

THE UNIVERSITY PRESS OF VIRGINIA
Copyright © 1979 by the Rector and Visitors
of the University of Virginia

First published 1979

Library of Congress Cataloging in Publication Data

Humm, Harold Judson.
  The marine algae of Virginia.

  (Special papers in marine science; no. 3)
  Bibliography: p.
  Includes index.
  I. Marine algae—Virginia—Identification. I. Title.
II. Series.
QK571.5.V8H85   589'.39'2147   78–16319
ISBN 0–8139–0701–2

Printed in the United States of America

CONTENTS

PREFACE

Knowledge of the marine benthic algae of Virginia has been a conspicuous hiatus in our understanding of the distribution of these plants along the Atlantic coast of North America until the 1970s. The algae of New England have been well documented since 1937 by William Randolph Taylor's *The Marine Algae of the Northeastern Coast of North America;* those of North Carolina since 1920 by W. D. Hoyt's *Marine Algae of Beaufort, N.C., and Adjacent Regions;* those of the coast from North Carolina southward to the West Indies since 1960 by Taylor's *Marine Algae of the Tropical and Subtropical Coasts of the Americas.*

Since Virginia was located between two relatively well known areas, my interest was aroused in the Virginia flora, and collections of algae in Virginia were begun in 1942 when I was associated with the War Production Board and engaged in a search along the Atlantic coast for adequate quantities of certain algae suitable as a source of agar, a critical war material. The survey report included an estimation of the abundance of *Gracilaria verrucosa* growing in the bays along the Eastern Shore of Virginia and Maryland, and general collections were made at the same time.

In 1946, I was invited to the Virginia Institute of Marine Science (the Virginia Fisheries Laboratory at that time) by the director, Dr. Nelson Marshall, to make collections of algae in the lower York River in the vicinity of the laboratory, then located in Yorktown. Subsequent visits to the laboratory and to other Virginia coastal areas provided additional records of the summer flora. In 1956 a valuable collection of Virginia algae made by Katrine deWitt, principally at Lynnhaven Inlet and Virginia Beach, was sent to the herbarium of Duke University. Miss deWitt's collection was made mostly from 1946 to 1955, following her study of marine algae under the late Dr. H. L. Blomquist of Duke University.

During 1961 and 1962 collections of algae from the vicinity of the laboratory were sent to me by Dr. Marvin Wass of the Institute faculty. In 1963, winter collections were sent by Michael Castagna, scientist-in-charge of the Eastern Shore Laboratory of the Institute from the general vicinity of Wachapreague. These collections added significantly to the records.

Invitations from Dr. William J. Hargis, Jr., Director of the Virginia Institute of Marine Science, to teach courses in marine algae at the Institute during the summers of 1962, 1963, 1966, and in October 1968, made possible additional collections and field data.

An opportunity to prepare a first draft of this work was provided when I was named the first Jacques Loeb Associate in Marine Biology at the Rockefeller University during the academic year 1959–60 (at that time the Rockefeller Institute), sponsored jointly by the Office of Naval Research and the University. I decided, however, that the incompleteness of the species list at that time was too great to justify publication.

A total of 174 species of marine algae are now known to occur in the inshore waters of Virginia in which the salinity is generally higher than about fifteen parts per thousand. While this total represents an overwhelming majority of the species that occur in these waters, there are additional species yet to be found and reported. I predict that the total will reach 200 or more and that most of those not yet known are winter species or species restricted to offshore waters on the continental shelf.

Special acknowledgment is made to Jane Davis for preparation of most of the drawings in this book; to Fred C. Biggs, Information Director of the Virginia Institute of Marine Science, for invaluable assistance in arranging for its publication, for the photographs, and assistance in many other ways; to Dr. Francis Drouet for the Latin diagnosis of *Ulothrix endospongialis* and for many helpful suggestions; to Dr. James Fiore for critical reading of a part of the manuscript.

Harold J. Humm
Department of Marine Science
University of South Florida
St. Petersburg 33701
November 1978

INTRODUCTION

None of the floristic works on the marine benthic algae of the Atlantic coast of North America include the flora of Virginia except casually. Until 1965, there were only a few published records of Virginia marine algae, although the majority of species present could be inferred from the work of Hoyt (1917–18) and Taylor (1957).

Aziz (1965), in his dissertation on *Acrochaetium* in Atlantic coast waters, reported four species of this genus in Virginia. Zaneveld and Barnes (1965) studied the seasonal behavior of twenty-nine species of red, brown, and green algae from Virginia, and Zaneveld (1966) recorded thirteen species (*sensu* Drouet and Daily, 1956; Drouet, 1968, 1973), of bluegreen algae commonly found in the lower Chesapeake Bay. Mangum, Santos, and Rhodes (1968), in a study of the marine annelid *Diopatra cupraea* (Bosc), collected the algae attached by this worm to the top of its tubes in the York River off Sandy Point near the Naval Weapons Station and listed nineteen species. Wulff and Webb (1969), in a study of the intertidal zonation of algae on pilings of the old ferry pier at Gloucester Point in the York River, recorded twenty-six species.

Rhodes (1970a) made summer, winter, and spring collections in Burtons Bay on the Eastern Shore of Virginia, where he found forty-three species of red, brown, and green algae, twenty-four of which were present during winter and spring and thirty-one during summer. Rhodes (1972) recorded *Porterinima fluviatile* and Rhodes and Connell (1973) showed that *Petalonia fascia* and *Scytosiphon lomentaria*, winter-spring species in Virginia, are present during the warmer months in the form of small crusts on oyster shells and other substrata. Rhodes (1976) added six more new records of brown algae to the Virginia flora. Ott (1973) compiled published lists and herbarium records of Virginia algae for a total of about 124 marine species of the four major groups. Zaneveld and Willis (1974, 1976) listed thirty-eight green and forty brown algal species from Virginia, respectively.

THE COASTAL WATERS OF VIRGINIA

The coastal waters of Virginia are, for the most part, divisible into three major areas with reference to marine algal habitats: (1) the Eastern Shore, characterized by extensive coastal bays and salt marshes from the Maryland state line to Cape Charles; (2) the Chesapeake Bay and Tidewater area, one of the largest estuaries in the world, opening to the sea between Cape Charles and Cape Henry; (3) the offshore waters of the continental shelf along the Eastern Shore and from Cape Henry to the North Carolina line. The latter area provides the least substrata for attachment of benthic algae. There are, however, a few wrecks, artificial reefs, and scattered shells in depths of fifteen to one hundred meters or more. In the offshore area benthic algae are to be found only on shells, on wrecks, or on sea buoys or other aids to navigation, as these constitute the only suitable substrata. South of the mouth of the Chesapeake Bay, from Cape Henry to the North Carolina border, there may be significant dilution of the shallow water over the continental shelf from water flowing out of the Chesapeake Bay, as the prevailing inshore currents in the area are southerly.

Salinity Range. In general, the waters of Virginia that support benthic marine algae are of somewhat reduced salinity in the bays and nearshore areas along the Eastern Shore, are of slightly to moderately reduced salinity in the lower Chesapeake Bay, and are of progressively reduced salinities up the bay and up the rivers of the lower part of the bay to essentially fresh water.

Bottom salinities in the lower Chesapeake Bay are generally somewhat higher than surface salinities as a result of bottom currents that move into the bay from the continental shelf, especially when the tide is rising. River discharge, after considerable mixing in the lower bay, tends to move out to sea at the surface or in the upper layers (Norcross, Massman, and Joseph, 1962).

At the mouth of the York River around Gloucester Point, the average annual fluctuation in salinity of the upper few meters of water is from about twelve to twenty-four parts per thousand, with the higher salinities in the spring and the lower salinities in the fall. During the more lengthy periods of high salinity, a marked invasion by many species of marine algae occurs from the lower bay toward the upper bay and into the river mouths. This phenomenon was quite striking during the spring and summer of 1966 and 1977, when rainfall was significantly below normal.

Temperature-related Ecology. Water temperature in the lower Chesapeake Bay ranges annually from a low in February of 2° to 5° C to a high in July and August of 26° to 28° C (fig. 1), if temperature data

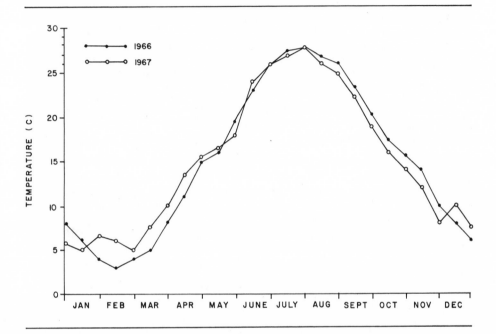

Fig. 1.   Water temperature at Gloucester Point, 1966 and 1967

recorded at the end of the laboratory pier of the Virginia Institute of Marine Science at Gloucester Point (McHugh, 1959) are applicable to other parts of the lower bay. In general, algae that are year-round in these waters are subjected to an annual range of more than 20° C, perhaps as much as 25° C.

The available data on water temperature, however, do not provide exact information on the water temperature of all algal habitats. The microclimate to which the plants are subjected remains unknown.

Algae in the intertidal zone are subjected both to the prevailing water temperature at high tide and to the prevailing air temperature at low tide. They must tolerate the most rapid temperature changes and also the greatest magnitude of temperature change that occurs in any algal habitat. In the case of intertidal rocks, the higher temperatures of the microclimate are significantly greater than any recorded air temperature of the locality because of heating of the rocks by infrared radiation. In the upper intertidal zone, a change in excess of 20° C may take place in a matter of minutes when the tide rises.

Even where precise temperature data are available, the effect of this major factor cannot be fully assessed. The influence of temperature is often mitigated or accentuated by other environmental factors that also vary. The effect of temperature may be more a result of the rate of change than of the magnitude of change. A rapid increase of 5° may be more harmful than a slow increase of 10° C. Some species may tolerate a water temperature change of 5° for a few hours, but not for several days. A low temperature with high salinity may be more tolerable than a low temperature with low salinity.

Setchell (1915) expressed the opinion that most marine algae grow actively over a temperature range of no more than 10° C and hence are confined to only one of his geographical zones of the world based upon temperature. For this reason, the marine algae of Virginia exhibit an annual alternation of warm-water and cold-water species, with the changeover occurring mostly during April and November. Even those species that are present the year around in the vegetative state may grow and reproduce only within a 10° temperature range. It is remarkable how few species have been studied to determine the temperature range of growth and the relationship between temperature and reproduction.

CHARACTERISTICS OF THE VIRGINIA FLORA

The marine waters of Virginia are located in the southern part of the range of those algae of cold-water affinities that occur from the region north of Cape Cod southward to the area of Beaufort, North Carolina; they are also in the upper part of the range of those species of tropical affinities that occur from the Caribbean Sea northward to Cape Cod (Humm, 1969).

Thus it appears that very few species reach either their northern or their southern limit along the coast of Virginia except in isolated or atypical habitats or as stragglers. There is no distinct transitional zone in the inshore waters of Virginia. All the red, brown, and green algae of Virginia are species that belong either to the center of distribution of the cold-water benthic flora north of Cape Cod or to the center of distribution of the tropical flora of the Caribbean Sea and West Indies. Most of the bluegreen algae do not seem to belong specifically to one of these centers of distribution, as they are remarkably eurythermal.

The majority of the algae of Virginia have cold-water affinities, as the coastline is nearer to the inshore southern boundary of the cold-water flora (Cape Cod) than it is to the inshore northern boundary of the tropical flora (Cape Canaveral, Florida). The algae between Cape Cod and Cape Canaveral are the more eurythermal species that belong either to the northern or to the tropical flora, as there appear to be no endemic species between these two principal transitional zones or boundaries.

One of the exceptions to the general rule of no range limits in Virginia is the tropical brown alga *Dictyota dichotoma*. This species has been observed to occur in good abundance each summer in Hummock Channel near Wachapreague (Humm, 1963a; Rhodes, 1970a). Beaufort, North Carolina, has been considered to be the northern limit of this species since the publication of Hoyt (1917–18) on the algae of the Beaufort area. In North Carolina, *Dictyota dichotoma* appears each spring about mid-April and disappears each fall about mid-November. The approximate dates of its annual appearance and disappearance in Virginia are unknown, but it seems likely that its season of vegetative growth there is slightly shorter than in North Carolina, unless the Virginia population represents a strain that tolerates lower water temperature than the populations farther south.

The discovery of *Dictyota dichotoma* on the Eastern Shore of Virginia suggests that others of the more eurythermal members of the tropical flora may also reach this area as a northern limit. However, the presence of *Dictyota dichotoma* in Virginia, and possibly of other species of tropical affinity not yet reported north of North Carolina, does not invalidate the concept of the Beaufort, North Carolina, area as one of the two major transition zones of the inshore waters of the Atlantic coast. The explanation would seem to be that there are a few algal habitats north of Cape Hatteras in which the water-temperature characteristics are more like those south of the Cape than like the surrounding waters, permitting a disjunct distribution. *Dictyota dichotoma* has reached this area and is able to survive the winter in it. A similar situation is exhibited by certain warm bays north of Cape Cod in which are found several species that reach their normal northern limit on the south side of Cape Cod in Buzzards Bay.

Dictyota dichotoma in the Wachapreague area and other similar habitats in Virginia evidently persists through the winter in the form of viable holdfasts or basal portions of plants. These are capable of sprouting in the spring when the water temperature becomes favorable. This is the type of survival through the unfavorable season shown by Rhodes and Connell (1973) to occur in *Petalonia fascia* and *Scytosiphon lomentaria* in Virginia, but at the opposite time of year. While dormant spores or zygotes of *Dictyota dichotoma* may also survive the winter in Virginia, the regular seasonal appearance of this species in Hummock Channel and elsewhere indicates that its presence does not depend upon an annual "re-seeding" of the area by spores or vegetative fragments of plants transported by currents from farther south each spring and summer.

A sporadic invasion of species from farther south may occur occasionally as a result of the reversal of inshore currents flowing along the Virginia coast, as recorded by Norcross, Massman, and Joseph (1962) during July 1960.

The only bluegreen alga known to be seasonal in Virginia waters is the strictly planktonic *Oscillatoria erythraea* (*Trichodesmium*) that appears each summer.

A sharp distinction cannot be made between the year-round group and those species more or less restricted to the colder months, as some species persist throughout the summer in varying degrees. Among year-round species, there is a period of most rapid vegetative growth and often a period of reproduction as well as a period of semidormancy, depending upon the season of the year to which each species is best adapted.

Since year-round collecting in Virginia waters has been limited, records are not complete and some published records are unreliable. In many cases, the year-round or seasonal nature of the species has been inferred from its known distribution and behavior outside Virginia.

A NOTE ABOUT KEYS

In constructing the keys in this book, the goal has been to make them as helpful as possible. However, a word of caution may be in order.

A key is simply a help in the process of determination of the name of a plant. It is not necessarily a direct means to that end. Descriptions, illustrations, and herbarium or preserved specimens that have been authentically determined should also be consulted. Students should not develop the impression, or be taught, that species determinations are merely a matter of "keying out."

An additional difficulty with local guides such as this one is the fact that they never contain all the species that occur in the area they encompass. Thus, when a student encounters a species not in the keys provided, he may be inclined to refer it to an incorrect name. Consultation of descriptions, illustrations, and reference specimens may prevent such an error.

SYSTEMATIC LIST OF SPECIES

CYANOPHYTA

Class Myxophyceae

  Coccogonales

   Chroococcaceae

   Agmenellum quadruplicatum
   Agmenellum thermale
   Anacystis aeruginosa
   Anacystis cyanea
   Anacystis dimidiata
   Anacystis marina
   Anacystis montana
   Coccochloris elabens
   Coccochloris peniocystis
   Coccochloris stagnina
   Gomphosphaeria aponina
   Johannesbaptistia pellucida

   Chamaesiphonaceae

   Entophysalis conferta
   Entophysalis deusta

  Hormogonales

   Oscillatoriaceae

   Arthrospira neapolitana
   Microcoleus lyngbyaceus

Microcoleus vaginatus
Oscillatoria erythraea
Oscillatoria lutea
Oscillatoria submembranacea
Porphyrosiphon notarisii
Schizothrix arenaria
Schizothrix calcicola
Schizothrix mexicana
Schizothrix tenerrima
Spirulina subsalsa

Nostocaceae

Anabaina oscillarioides
Calothrix crustacea
Calothrix parietina
Nostoc spumigena
Scytonema hofmannii

Stigonemataceae

Brachytrichia quoyi
Mastigocoleus testarum

RHODOPHYTA

Class Rhodophyceae

Subclass Bangiophycidae

Goniotrichales

Goniotrichaceae

Asterocytis ramosa
Erythrocladia subintegra
Erythrotrichia carnea
Erythrotrichia rhizoidea
Goniotrichum alsidii

Bangiales

Bangiaceae

Bangia fuscopurpurea
Porphyra leucosticta
Porphyra miniata
Porphyra umbilicalis

Subclass Florideophycidae

Nemaliales

Acrochaetiaceae

Acrochaetium alcyonidii
Acrochaetium dasyae
Acrochaetium daviesii
Acrochaetium flexuosum
Acrochaetium robustum
Acrochaetium sagraeanum
Acrochaetium thuretii
Acrochaetium trifilum
Acrochaetium virgatulum

Gelidiales

Gelidiaceae

Gelidium crinale

Cryptonemiales

Corallinaceae

Corallina officinalis
Fosliella farinosa
Fosliella lejolisii
Jania capillacea
Melobesia membranacea

Gigartinales

Solieriaceae

Solieria tenera

Hypneaceae

Hypnea musciformis

Gracilariaceae

Gracilaria foliifera
Gracilaria verrucosa

Phyllophoraceae

Gymnogongrus griffithsiae

Rhodymeniales

Champiaceae

Champia parvula
Lomentaria baileyana

Ceramiales

Ceramiaceae

Antithamnion cruciatum
Callithamnion baileyi
Callithamnion byssoides
Callithamnion corymbosum
Callithamnion roseum
Ceramium diaphanum
Ceramium fastigiatum
Ceramium rubriforme
Ceramium rubrum
Ceramium strictum
Griffithsia tenuis
Spermothamnion turneri
Spyridia filamentosa
Trailliella intricata

Delesseriaceae

Caloglossa leprieurii
Grinnellia americana

Dasyaceae

Dasya baillouviana

Rhodomelaceae

Bostrichia rivularis
Chondria baileyana
Chondria dasyphylla
Chondria sedifolia
Chondria tenuissima
Polysiphonia denudata
Polysiphonia harveyi
Polysiphonia nigrescens
Polysiphonia subtilissima

XANTHOPHYTA

Class Xanthophyceae

Vaucheriales

Phyllosiphonaceae

Ostreobium quekettii

Vaucheriaceae

Vaucheria thuretii

PHAEOPHYTA

Class Phaeophyceae

Ectocarpales

Ectocarpaceae

Ectocarpus dasycarpus
Ectocarpus elachistaeformis
Ectocarpus intermedius
Ectocarpus penicillatus
Ectocarpus siliculosus
Ectocarpus tomentosus
Giffordia indica
Giffordia mitchelliae
Pylaiella littoralis
Sorocarpus micromorus
Streblonema oligosporum

Sphacelariales

Sphacelariaceae

Sphacelaria cirrosa
Sphacelaria furcigera
Sphacelaria fusca

Dictyotales

Dictyotaceae

Dictyota dichotoma

Chordariales

Myrionemataceae

Ascocyclus magnusii
Myrionema corunnae
Myrionema strangulans

Ralfsiaceae

Ralfsia verrucosa

Elachistaceae

Elachista fucicola

Corynophlaeaceae

Leathesia difformis

Chordariaceae

Cladosiphon zosterae

Stilophoraceae

Stilophora rhizodes

Desmarestiales

Arthrocladiaceae

Arthrocladia villosa

Dictyosiphonales

Striariaceae

Hummia onusta
Stictyosiphon soriferus
Striaria attenuata

Punctariaceae

Asperococcus fistulosus
Desmotrichum undulatum
Petalonia fascia
Petalonia zosterifolia
Punctaria latifolia
Punctaria plantaginea
Scytosiphon lomentaria

Dictyosiphonaceae

Dictyosiphon eckmani
Dictyosiphon foeniculaceus

Fucales

Fucaceae

Ascophyllum nodosum
Fucus edentatus
Fucus vesiculosus

Sargassaceae

Sargassum filipendula
Sargassum fluitans
Sargassum natans

CHLOROPHYTA

Class Chlorophyceae

Ulotrichales

Ulotrichaceae

Ulothrix endospongialis
Ulothrix flacca
Ulothrix subflaccida

Chaetophoraceae

Entocladia viridis
Entocladia wittrockii
Phaeophila dendroides
Pilinia rimosa
Protoderma marinum

Gomontiaceae

Gomontia polyrhizum

Ulvales

Ulvaceae

Enteromorpha clathrata
Enteromorpha compressa
Enteromorpha erecta
Enteromorpha intestinalis
Enteromorpha lingulata
Enteromorpha linza
Enteromorpha marginata
Enteromorpha minima
Enteromorpha plumosa
Enteromorpha prolifera
Enteromorpha torta
Monostroma leptodermum
Monostroma oxyspermum
Percursaria percursa
Ulva curvata
Ulva rotundata

Cladophorales

Cladophoraceae

Chaetomorpha aerea
Chaetomorpha linum

Cladophora albida
Cladophora crystallina
Cladophora expansa
Cladophora laetevirens
Cladophora rupestris
Cladophora sericea
Rhizoclonium kerneri
Rhizoclonium kochianum
Rhizoclonium riparium
Rhizoclonium tortuosum
Urospora mirabilis

Siphonales

Bryopsidaceae

Bryopsis hypnoides
Bryopsis plumosa

CYANOPHYTA

Bluegreen algae exhibit the simplest cellular organization of any of the algae. They are the most primitive plants on the earth. Their procaryotic cell is internally separable into a central, colorless area in which the nuclear proteins are concentrated and an outer area in which the pigments are distributed. These two areas are not separated by a membrane. There is no nucleus, in the usual sense, and no chromatophores or chloroplasts. There is no known sexual reproduction, although a gene-exchange process apparently does occur, such as that known in the Eubacteriales (Carr and Whitton, 1973, p. 212). Cell division is a simple cleavage of the cell by formation of a membrane that grows inward from the periphery. There is no mitosis, no flagellated cells.

Bluegreen algal cells secrete polysaccharides or mucopolysaccharides that usually tend to accumulate around the cells and form a gelatinous envelope or a distinct sheath. The thickness and the consistency of this sheath vary greatly with environmental conditions, and a lack of knowledge of this fact has led to the description of scores of genera and hundreds of species based largely upon sheath differences. Drouet and Daily (1956) and Drouet (1968, 1973, 1978) have shown, partly by culturing, the influence of the environment upon sheath formation and have revised the classification on the basis of protoplast characters, believed to be genetically controlled to a greater extent and hence more reliable.

Among coccoid species, the sheath is responsible for the formation of colonies of cells, because of its tendency to hold them in groups. In filamentous species, the trichomes may move forward or backward within the sheath, and the sheath often contains a trichome that has fragmented into sections of various lengths. Trichome segments often emerge from the open end of a sheath, and when they do, this section of a trichome has been referred to as a hormogone. Hormogones may become temporarily planktonic and may then serve to disseminate the species. The term filament, when applied to bluegreen algae, includes both trichome and sheath. Diameter measurements are usually given for the trichome, as these are more useful in taxonomy than the diameter of the filament.

Under some conditions, the sheath may become variously pigmented. As with coccoid species, the degree of development of the sheath in filamentous species also determines the extent and nature of "colony" formation. The colony or aggregation of filaments may take the form of tufts or layers of filaments forming strata. Some filamentous species may be completely embedded in a common gelatinous matrix, some partially so.

The sheath material of at least some bluegreen algae is similar in physical properties to agar or carrageenan, cell-wall polysaccharides of certain red algae. The sheath may be soluble in boiling water and, upon cooling, may form a thermally reversible gel.

Variation in physical properties of the sheath—including firmness, degree of hydration, and diffluence—may be related in variation in proportion of several sheath polysaccharide constituents, the production of which in turn varies with environmental conditions. Further studies of the chemistry and properties of bluegreen extracellular polysaccharides are needed.

The photosynthetic pigments of the bluegreen algae occur in the peripheral region of the cell in double-layered lamellae and are not localized in chromatophores. Only chlorophyll *a* is present in the bluegreens, but of equal importance in photosynthesis are three phycobilin (biliprotein) pigments, allophycocyanin (blue), phycocyanin (blue), and phycoerythrin (red). Light energy absorbed by these pigments is passed on to chlorophyll *a* with high efficiency, and chlorophyll *a* carries on phosphorylation. The red and blue pigments absorb wavelengths of light not readily absorbed by chlorophyll.

The relative abundance of the various pigments, including the carotenes and xanthophylls, in the bluegreen algae is much influenced by environmental conditions, especially the intensity, quality, and regime of light. Consequently, the gross color of any species is subject to much variation, and color is not a reliable taxonomic character.

Myxophycean starch is the principal stored food of bluegreen algae. It is similar to floridean starch of red algae but differs in chain length (Meeuse, 1962). It occurs principally as submicroscopic granules. In addition, bluegreen algae store polyphosphates as microscopically visible granules, referred to in the past as volutin, metachromatic granules, or cyanophycin. They are often conspicuous as refringent granules against the end walls of certain filamentous species and have been used as a taxonomic character (Drouet, 1968).

Although flagella are unknown among bluegreen algae, many filamentous species exhibit an oscillating or creeping motility when in contact with a solid or semisolid surface, rotating on their long axes as they move. A number of tentative explanations for this movement have been proposed, including the directional secretion of polysaccharide through pores, followed by rapid hydration with a propelling effect. A recent and highly plausible explanation is that of Jarosch (1962) and of Halfen and Castenholz (1971), who account for this movement on the basis of the presence in the cell wall of numerous fibrils arranged in parallel array in a sixty degree helical orientation. A rapid succession of unidirectional waves is propagated along these fibrils, propelling the trichome along a solid or semisolid surface and also causing it to rotate. These fibrils are similar in size and structure to the flagella of bacteria. Castenholz (chap. 15 in Carr and Whitton, 1973) reviews the subject in detail.

There is no known sexual reproduction among bluegreens. So-called spores are produced by certain coccoid species. "Endospores" result from the repeated internal division of a large cell into many small vegetative cells that may be retained within the parent cell sheath for a time in the genus *Entophysalis*. Among filamentous genera, fragmentation of the trichome into short segments referred to as hormogonia or "hormospores" is an important dissemination mechanism. Neither *Entophysalis* endospores nor hormospores are true spores.

The only distinct spores of the bluegreens are produced in the families Nostocaceae and Scytonemataceae. A vegetative cell enlarges; it becomes dense with pigments and food reserves; the wall thickens, and the sheath may become continuous around the ends of the cell, followed by cessation of sheath production and severing of cytoplasmic connections with adjacent cells as the plasmodesmata become plugged. This type of spore has been referred to as an akinete or a conidium. Apparently it possesses somewhat greater resistance to adverse conditions than do vegetative cells.

Spores are of common occurrence in freshwater species of the genera that produce them but are rare among marine species. The greater stability of the marine environment probably does not favor their development.

The most distinctive modification of a vegetative cell among the bluegreens is the heterocyst, found in all filamentous bluegreens except for the large family Oscillatoriaceae. Heterocysts, in developing from a vegetative cell, first secrete a new, three-layered outer wall, followed by partial loss of pigments and of particulate or granular material. During this development, cytoplasmic connections become established with adjacent cells. As the heterocyst matures, a collar of polysaccharide material develops on the heterocyst side of the wall at the terminus of the plasmodesma. These pluglike structures, as well as the yellowish color and translucency, help to identify a heterocyst (Fritsch, 1951; Lang and Fay, 1971).

From the time of its discovery until recently, the function of the heterocyst has been a puzzle. It is now known to be the principal site of nitrogen fixation. Since it does not carry on the photosynthetic processes that involve oxygen evolution, it provides a reduced microenvironment in which the enzyme nitrogenase can function efficiently (Stewart, Haystead, and Pearson, 1969).

Bluegreen algae are more abundant and of greater variety on land and in fresh water than in the sea. Nevertheless, they constitute an important group of marine plant life in all parts of the sea in which photosynthesis occurs, but especially in the intertidal zone. Rocks, seawalls, pilings—the upper intertidal zone in general—are often clearly marked by a mixture of bluegreens. Along the Virginia coastline, they typically form a high intertidal black zone dominated by *Calothrix crustacea*. On creosoted pilings, the *Lyngbya confervoides* ecophene of *Microcoleus lyngbyaceus* is usually the dominant or the only alga because of its tolerance of the toxic compounds in treated pilings. In general, the intertidal bluegreens are remarkably tolerant of wide and sudden environmental changes.

Bluegreens are also found below the intertidal zone, especially as epiphytes on larger algae and seagrasses, but also on invertebrates and on stones and shells. In benthic habitats they are easily overlooked. One bluegreen species, *Oscillatoria erythraea*, is strictly planktonic, but it is mostly limited to warm waters.

At least five species of bluegreen algae regularly penetrate shells and other forms of limestone—only one of which, *Mastigocoleus testarum*, is found exclusively within limestone. Their presence within this substrate is usually indicated by a greenish surface tinge. They can be studied only after pieces of the substrate are decalcified in dilute hydrochloric acid. This treatment does not appear to alter the morphology of the cells or filaments. Along the coastline of Virginia and throughout the Chesapeake Bay, old shells exposed to light in the intertidal zone almost invariably contain one to several species. They are also found in the tests of calcareous bryozoans and in the calcareous tubes produced by certain annelid worms.

Perkins and Tsentas (1976) studied the rate of penetration of limestone by bluegreen algae at St. Croix, Virgin Islands, by placing pieces of limestone lacking algae at various depths from the intertidal zone to thirty meters and examining these experimental substrates at regular intervals. Most of the pieces were infested with limestone-boring bluegreens within a few weeks and were heavily colonized within six months at all depths to thirty meters.

The distinctiveness of the bluegreen algae is derived principally from the fact that they are procaryotic like the bacteria, features associated with primitiveness or with early forms of life. The characteristics they have in common with bacteria have led to a growing acceptance of the inclusion of bacteria and bluegreen algae in the same kingdom or in the same phylum or division. In the seventh edition of Bergey's Manual of Determinative Bacteriology (Breed, Murray, and Smith, 1957), the bluegreen algae comprise the class Schizophyceae in the division Protophyta. In the eighth edition (Buchanan and Gibbons, 1974), the bluegreens are raised to phylum level, as the kingdom Procaryotae is divided into two divisions or phyla, the Cyanobacteria and the Bacteria.

The bluegreen algae are, in fact, bluegreen bacteria, as they differ from certain nonphotosynthetic bacteria only in their possession of chlorophyll *a*. Because of their role and distribution in the environment, they will always be treated with the algae. Whether they are referred to as "algae" or "bacteria" is of minor importance. Margulis (1968), in one of the most logical classifications yet proposed, places them in the kingdom Monera.

Baker and Bold (1970) have carried on extensive and critical studies of seven species (*sensu* Drouet) of the family Oscillatoriaceae, five of which occur commonly in the sea. They employed unialgal or axenic cultures obtained from a wide variety of sources and studied them with the goal of evaluating the principal taxonomic criteria used by Drouet (1968) in his revision of this family.

They found cell granulation to be a variable character, especially in the number of granules and their arrangement. Since these granules represent stored nutrients, they vary with the availability of these nutrients in the environment. The morphology of mature tip cells was more constant in culture. The sheath, they observed, was as reliable and constant as the morphology of the tip cells and was more reliable than the arrangement of granules in the cells.

Stanier et al. (1971) have also conducted extensive studies of bluegreen algae in culture. Their work with the Chroococcaceae has led them to the conclusion that species in this family cannot be characterized adequately on the basis of morphology and that "all future taxonomic work on the Chroococcales should be based on the isolation and comparison of pure strains."

An excellent study of the bluegreen algae of Delaware is Ralph's (1977) "The Myxophyceae of the marshes of southern Delaware," a work that applies almost equally well to the coastal marshes of Virginia.

CLASS MYXOPHYCEAE

Bluegreen algae exhibit so little variation of fundamental morphological characters that they are all placed in one class, the Myxophyceae. In
some classifications, this class has been the Cyanophyceae, but in the system used here the stem of the latter is used as the phylum name, Cyanophyta.

Division of the class Myxophyceae into more than one order would
appear to be more a matter of convenience than of a sound and logical
recognition of differences of ordinal magnitude. West and Fritsch
(1927), Lindstedt (1943), and Smith (1950) recognized three orders.
Tilden (1910) and Newton (1931) recognized only two orders, Coccogonales and Hormogonales. The latter classification is followed here
because of its optimum convenience.

Key to the Orders of the Class Myxophyceae

Cells essentially coccoid, the plants unicellular or held together in colonies by a gelatinous sheath; not filamentous (except for *Entophysalis
deusta* that has penetrated limestone and the pseudofilamentous cell
rows in *Johannesbaptistia*) . . . . Coccogonales

Plants multicellular and filamentous, that is, end walls of adjacent cells
not separated by sheath material . . . . Hormogonales (p. 33)

ORDER COCCOGONALES

Plants in the form of single cells or colonies in which the cells are held
together by a gelatinous sheath secreted from all surfaces of the cell.
Reproduction is by simple cell division into two daughter cells of equal
or unequal size. In the genus *Entophysalis* of the family Chamaesiphonaceae, however, large cells sometimes divide simultaneously into a few
to many small daughter cells that are retained within the original sheath
for a time. These are referred to as endospores. Dissemination occurs
as a result of separation of cells or groups of cells from colonies. Attachment of colonies or cells to solid surfaces is a function of the adhe-
sive nature of the extracellular polysaccharides.

The coccoid Myxophyceae are treated here in accordance with the
admirable revision of this group by Drouet and Daily (1956). Taxonomy of this group before the publication of the Drouet and Daily monograph, or by authors who have not chosen to follow that work, may
involve hundreds of genera and species relegated into synonymy by
Drouet and Daily. They have shown that the trivial morphological

characters upon which these numerous invalid genera are based are simply variations resulting from environmental influences and cannot serve, in themselves, to distinguish genera or species.

Drouet and Daily (1956) recognize three families, one of which, the Clastidiaceae, occurs only in fresh water and is not treated here.

Key to the Families of the Order Coccogonales

Plants unicellular or in colonies, planktonic or attached; cell division results in two equal daughter cells . . . . . . Chroococcaceae

Plants at first unicellular, becoming attached and producing a stratum or cushion of cells from which filamentlike rows may penetrate the substratum; cell division produces unequal daughter cells, the apical one smaller; large cells may divide internally to form many "endospores" . . . . Chamaesiphonaceae (p. 30)

FAMILY CHROOCOCCACEAE

Members of the Chroococcaceae may occur in the plankton or may be loosely attached to solid substrata or to other algae, submerged aquatics, or animals. They often occur in abundance on sand or muddy sand that is not much disturbed by waves or currents. They may be attached to the sand grains or loose among them.

Key to the Genera of the Family Chroococcaceae

The keys that follow include only those species known to occur or expected to occur along the coast of Virginia in waters having a salinity of about fifteen parts per thousand or greater.

1. Cells prior to division ovoid to cylindrical; dividing in a plane perpendicular to the long axis . . . . . *Coccochloris* (p. 22)

1. Cells prior to division spherical, ovoid, cylindrical, or pyriform; not dividing in a plane perpendicular to the long axis 2

2. Cells disk-shaped prior to division and arranged in a single linear series resembling a filament, but each cell separated by polysaccharide sheath material . . . . . *Johannesbaptistia* (p. 28)

2. Cells not in a single linear series 3

3. Cells usually pyriform, radially arranged in a spherical gelatinous matrix . . . . *Gomphosphaeria* (p. 30)

3. Cells not radially arranged in a gelatinous matrix 4

4. Colonies a flat plate; the cells in regular rows . . . .
. . . . *Agmenellum* (p. 29)

4. Colonies spherical or irregular in shape; cells spherical before division and usually irregularly distributed in the gelatinous matrix . .
. . . . *Anacystis* (p. 24)

GENUS COCCOCHLORIS

Plants unicellular or in many-celled colonies, the cells subspherical to cylindrical (mature cells always elongate), dividing in one plane only, perpendicular to the long axis of the cell.

Key to the Species of the Genus Coccochloris

1. Cells 4–8 μm in diameter, to three times as long as the diameter
. . . . *C.   stagnina*

1. Cells relatively longer 2

2. Cells 1–3 μm in diameter, to twelve times as long . . . . . .
. . . . *C. peniocystis* (p. 23)

2. Cells 2–6 μm in diameter, to eight times longer than the diameter
. . . . *C. elabens* (p. 23)

COCCOCHLORIS STAGNINA SPRENGEL

(Fig. 2)

Colonies vary in color, depending upon the environment; usually of many cells, occurring in aerial and subaerial situations as well as in fresh water and sometimes in brackish and marine waters. Cells with blunt ends, 4–8 μm in diameter and to 3 diameters long, the longer cells usually in the process of division.

Known from a number of freshwater localities in Virginia (Drouet and Daily, 1956, p. 22) and probably common in the Chesapeake Bay. Of worldwide distribution.

Sprengel, 1807; Drouet and Daily, 1956, p. 15, figs. 145–63; Cocke, 1967, p. 5, figs. 2–8.

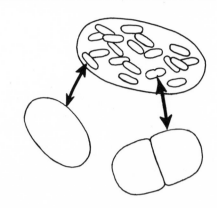

Fig. 2. *Coccochloris stagnina:* at left, a small colony in a gelatinous matrix; upper right, products of a recently divided cell that have not separated; lower right, a single cell before division

COCCOCHLORIS PENIOCYSTIS DROUET AND DAILY

Colonies blue-green or variously colored (pale violet, rose, red-brown), depending upon the environment. Cells 1–3 μm in diameter, to 12 diameters long, usually curved, the ends rounded or attenuate to conical.

Intertidal and below, on sand or adhering to oyster shells, pilings, or other solid objects in the Chesapeake Bay. This species is primarily a freshwater one, but it is also found in brackish water along the length of both the Atlantic and Pacific coasts of North America.

Kützing, 1845–71, 1:25, pl. 36, fig. 7 (as *Gloeocapsa peniocystis*); Howe, 1920, p. 620 (as *Gloeothece rupestris* [Lyngbye] Bornet); Drouet and Daily, 1956, p. 31, figs. 170–72 (with complete synonymy); Aziz and Humm, 1962, p. 56; Cocke, 1967, p. 7, figs. 11–12.

COCCOCHLORIS ELABENS DROUET AND DAILY

(Fig. 3)

Colonies usually yellow-brown, olive, or violet, of a few to many cells. Cells typically cylindrical or elliptic-cylindrical, straight, the ends rounded-truncate; 2–6 μm in diameter and several to 8 diameters long.

In brackish and marine habitats this species is often found in moist intertidal sand that is not much disturbed by waves or currents.

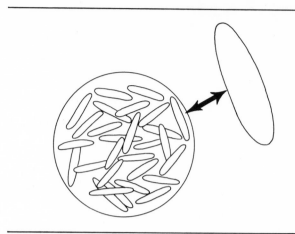

Fig. 3. *Coccochloris elabens:* at left, a small colony; right, a single cell, enlarged

Though apparently not yet recorded for Virginia, it is known from many stations in New England, from a marsh pool at Chance, Somerset County, Maryland, and from North Carolina, South Carolina, and Florida.

Drouet and Daily, 1948, p. 77; Drouet and Daily, 1956, p. 28, figs. 164–69; Aziz and Humm, 1962, p. 56; Cocke, 1967, p. 6, figs. 9–10.

GENUS *ANACYSTIS*

Plants in the form of colonies of spherical cells produced by successive divisions in three planes perpendicular to each other. The cells usually become irregularly distributed in the gelatinous matrix but may be distributed in regular rows in a three-dimensional arrangement.

Anacystis thermalis (Meneghini) Drouet and Daily, especially *forma thermalis* Drouet and Daily, has been reported from a freshwater station in Virginia (Drouet and Daily, 1956, p. 80). Although it is predominantly a freshwater species, it is known from marine habitats in Florida and surely occurs in the upper Chesapeake Bay and in the lower parts of the rivers that enter the bay. *Forma major* (Lagerheim) Drouet and Daily is also to be expected. It is known from Hargrove's Pond, New Kent County, Virginia (Drouet and Daily, 1956, p. 84).

Key to the Species of the Genus *Anacystis*

1. Cells with pseudovacuoles; plants developing as water blooms . .
    . . . . *A. cyanea* (p. 25)

1. Without pseudovacuoles; not producing water blooms . . . 2

2. Cells 0.5–2.0 μm in diameter . . . . . . *A. marina* (p. 26)

2. Cells larger 3

3. Cells mostly 3–6 μm in diameter . . . . *A. montana* (p. 26)

3. Cells mostly more than 6 μm in diameter 4

4. Cells mostly 6–12 μm in diameter, becoming spherical soon after division . . . . *A. aeruginosa* (p. 27)

4. Cells mostly 12–50 μm in diameter, the adjacent faces remaining flattened after division . . . . . . *A. dimidiata* (p. 27)

ANACYSTIS CYANEA DROUET AND DAILY

(Fig. 4)

Typically in the plankton of freshwater ponds, lakes, streams, and often forming water blooms during the warm season of the year. Cells with prominent pseudovacuoles, mostly 3–7 μm in diameter, and in a common gelatinous matrix.

Recorded from Queens Creek, York County, Virginia. Probably common at times in the Chesapeake Bay opposite river mouths. Of worldwide distribution.

Drouet and Daily, 1956, p. 36, figs. 1–11; Cocke, 1967, p. 8, figs. 14–15.

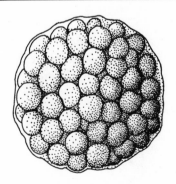

Fig. 4. *Anacystis cyanea,* a colony enclosed in a common gelatinous matrix

ANACYSTIS MARINA DROUET AND DAILY
(Fig. 5)

Cells spherical to ovoid, 0.5–2.0 μm in diameter, the colonies producing gelatinous strata on rocks or wood or mixed with other bluegreens in

Fig. 5.   *Anacystis marina*

small groups, especially in brackish water and in protected or marshy areas but also in full seawater. Vigorous development of this species occurred in a wooden vat on the pier of the Virginia Institute of Marine Science at Gloucester Point. Originally this vat was used to keep living fish or shellfish, but during the summer of 1966 it was not being used, and a heavy growth of *Schizothrix calcicola* developed, with many colonies of *A. marina* within its diffluent sheath material. The species is worldwide in distribution but has been recorded under several generic and many specific names (Drouet and Daily, 1956, p. 44).

Cocke, 1967, p. 9, fig. 16.

ANACYSTIS MONTANA (LIGHTFOOT) DROUET AND DAILY
(Fig. 6)

Plants usually in gelatinous strata on submerged or intertidal objects, sometimes in the plankton. The color is variable, especially where exposed to the air. The cells are 2–6 μm in diameter, irregularly distrib-

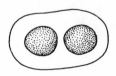

Fig. 6.   *Anacystis montana,* a two-celled colony

uted in the gelatinous matrix. Red or blue pigments are sometimes present in the extracellular polysaccharide and are reversible in response to changes in pH.

Drouet and Daily (1956) recognize three forms on the basis of cell size and sheath characters.

A. montana forma minor Drouet and Daily has been cultured from plankton obtained from the York River near Gloucester Point. It is of worldwide distribution.

Lightfoot, 1777 (as *Ulva montana*) ; Drouet and Daily, 1956, p. 45, figs. 16–90A; Cocke, 1967, p. 9, fig. 27.

ANACYSTIS AERUGINOSA DROUET AND DAILY

(Fig. 7)

Plants forming gelatinous masses on various submerged objects in brackish and quiet marine waters; those from deeper waters tend to be

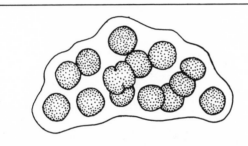

Fig. 7.   *Anacystis aeruginosa,* a small colony

rose or maroon in color. Cells 6–12 μm in diameter, spherical, in groups within a gelatinous matrix.

On stones, woodwork, and other algae in the York River at Gloucester Point, Virginia, and probably common throughout the Chesapeake Bay. Of worldwide distribution.

Drouet and Daily, 1948; Drouet and Daily, 1956, p. 76, figs. 108–13; Humm and Caylor, 1957, p. 232, pl. 1, fig. 1; Humm and Hildebrand, 1962, p. 234.

ANACYSTIS DIMIDIATA DROUET AND DAILY

(Fig. 8)

Plants as single cells or in colonies of two, four, or eight cells held together by the gelatinous sheath, the adjacent faces of the cells remain-

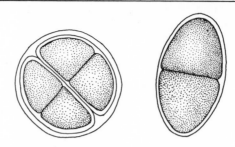

Fig. 8. *Anacystis dimidiata*

ing flattened until they escape from the sheath. Cells typically dark blue-green, 12–50 μm in diameter. In shallow fresh, brackish, and marine waters, rarely ever forming macroscopically visible groups.

Occasional on pilings and on other algae in the York River at Gloucester Point, Virginia, and probably common throughout the Chesapeake Bay. Of worldwide distribution.

Drouet and Daily, 1956, p. 70, figs. 100–107; Cocke, 1967, p. 12, figs. 28–31.

GENUS *JOHANNESBAPTISTIA*

Plants producing filaments of cells in which the cells are separated from each other by sheath material. Cells disk-shaped (shorter than wide), 3–20 μm in diameter, uniseriate in the filaments. Cell division is restricted to one plane through the diameter of the cell. One species.

JOHANNESBAPTISTIA PELLUCIDA TAYLOR AND DROUET

(Fig. 9)

Plants occur in shallow, quiet, fresh, brackish, and marine waters, often in mixtures of algae in marsh pools or tide pools.

While not yet recorded from Virginia, the species is known from a salt marsh near Crisfield, Maryland, and is to be expected throughout the Chesapeake Bay area.

Drouet, 1938; Drouet and Daily, 1956, p. 84, figs. 182–84; Cocke, 1967, p. 16, figs. 40–41.

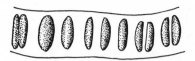

Fig. 9. *Johannesbaptistia pellucida,* a portion of a uniseriate colony or pseudofilament

GENUS *AGMENELLUM*

Plants in the form of a flat gelatinous plate in which the cells are in one layer and in a series of rows in two directions at right angles to each other as a result of cell division occurring in two planes only, alternately. (Synonym: *Merismopedia.*)

Key to the Species of the Genus *Agmenellum*

Cells 1.0–3.5 μm in diameter; colonies of up to 256 cells . . . . .
. . . . *A. quadruplicatum*

Cells 4–10 μm in diameter; colonies larger . . . . . *A. thermale*

AGMENELLUM QUADRUPLICATUM BREBISSON

Colonies flat, rectangular, blue-green or sometimes violet to rose in color. After division, the cells are 1.0–3.5 μm in diameter, closely or loosely arranged in the gelatinous matrix.

In intertidal sand of protected beaches, Chesapeake Bay. Of world-wide distribution.

Tilden, 1910, p. 43, pl. 2, fig. 35 (as *Merismopedium glaucum* [Ehrenberg] Naegeli); Drouet and Daily, 1956, p. 86, fig. 133; Cocke, 1967, p. 17, figs. 42–43.

AGMENELLUM THERMALE (KÜTZING) DROUET AND DAILY

(Fig. 10)

Cells 4–10 μm in diameter; colonies in shallow fresh, brackish, or salt waters or on sand or mud in protected places in the intertidal zone.

In intertidal sand near the laboratory pier, Virginia Institute of Ma-

Fig. 10. *Agmenellum thermale*

rine Science, Gloucester Point. Widely distributed in the Cheasapeake Bay; worldwide.

Kützing, 1843, p. 162 (as *Merismopedia thermalis*); Drouet and Daily, 1956, p. 89, figs. 134–42; Cocke, 1967, p. 18, figs. 44–48.

GENUS *GOMPHOSPHAERIA*

Plants in the form of a spherical gelatinous matrix in which the cells are radially arranged in a single peripheral layer. Cell division takes place in two planes perpendicular to each other. Individual sheaths are often visible around each cell, with the old portions of these sheaths in the form of a dichotomously branched structure radiating from the center of the matrix to the inner end of each cell. The cells are often pyriform, with the tapered end toward the center.

GOMPHOSPHAERIA APONINA KÜTZING

(Fig. 11)

Cells typically pyriform, 4–15 μm in diameter, the plants occurring in fresh, brackish, or marine waters, usually as plankton.

Found in tide pools along the York River at Yorktown, Virginia. Probably widely distributed in the Chesapeake Bay in similar habitats and in salt marshes. Of worldwide distribution.

Tilden, 1910, p. 38, pl. 2, figs. 23–28; Drouet and Daily, 1956, p. 98, figs. 178–80; Cocke, 1967, p. 20, figs. 53–54.

FAMILY CHAMAESIPHONACEAE

Plants unicellular to multicellular, eventually forming cushions or strata from which filaments of cells usually penetrate the substratum. Single

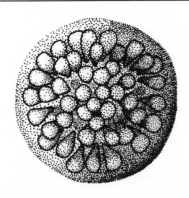

Fig. 11.   *Gomphosphaeria aponina*

cells are basally attached to the substratum by the gelatinous sheath. Cell division occurs at right angles to the long axis of the cell, or obliquely, and is unequal. The upper daughter cell is usually smaller. The upper cell may pass out of the sheath, or it may remain in place and divide again, in which case a cushion of cells develops. Occasionally a cell may divide repeatedly, producing an indefinite number of small cells known as endospores. All species are aquatic and do not tolerate prolonged exposure to the air, though the marine species are often in-tertidal. Drouet and Daily (1956) have reduced the family to one genus and five species from the scores of genera and hundreds of species based upon trivial characters that vary with the environment.

GENUS *ENTOPHYSALIS*

With characters of the family.

Key to the Genus *Entophysalis*

On rocks, wood, shells, and producing filaments of cells that penetrate wood and limestone . . . . *E. deusta*

On larger algae and living animals, generally not penetrating the sub-strate. . . . . *E. conferta*

ENTOPHYSALIS DEUSTA DROUET AND DAILY

(Fig. 12)

On rocks, wood, shells, and other solid substrata in brackish water and in seawater, intertidal or below, and often producing filamentous

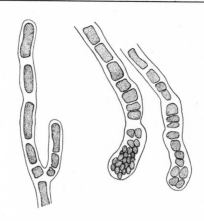

Fig. 12. *Entophysalis deusta:* pseudofilaments from within the calcium carbonate of a mollusc shell; the lower ends of those on the
right were at the surface, the middle filament having produced endospores.

rows of cells when a calcareous substratum is penetrated. Cells on the
surface usually form a cushion or stratum consisting of coccoid cells of
variable size, mostly 3–10 μm in diameter.

Abundant throughout the intertidal zone and below in the Chesapeake Bay and along the Eastern Shore of Virginia. Worldwide in distribution.

Drouet and Daily, 1956, p. 103, figs. 185–94, 247–50; Zaneveld,
1966, p. 113; Cocke, 1967, p. 23, figs. 55–56.

ENTOPHYSALIS CONFERTA DROUET AND DAILY
(Fig. 13)

On larger algae and living marine animals, sometimes penetrating the
polysaccharide of the algal host or the nonliving part of the animal
host. Plants may occur in the form of single cells, few- to many-celled
colonies, or as cushions or strata. Single cells may reach 20 μm in diameter, but those in strata are usually 3–6 μm in diameter. Endospores
are commonly produced, the endosporangia sometimes to 50 μm in
diameter.

Abundant in the Chesapeake Bay and along the Eastern Shore of
Virginia in the intertidal zone, but more commonly below low tide. Of
worldwide distribution.

Drouet and Daily, 1956, p. 111, figs. 196–215; Humm and Caylor,
1957, p. 234, pl. 1, figs. 6, 7; Cocke, 1967, p. 24, figs. 57–58.

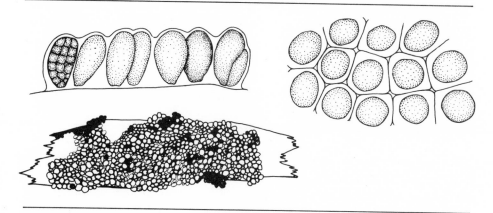

Fig. 13. *Entophysalis conferta:* lower left, a colony on the stem of the marsh grass, *Spartina;* upper left, lateral view of a group of cells; upper right, a group of cells in surface view. The cell in the upper left has produced endospores.

ORDER HORMOGONALES

The filamentous bluegreen algae produce multicellular plants in the form of filaments that may be attached or loose, unbranched, with false branches, or with true branches. The cells in a trichome are not separated from each other by polysaccharide sheath material, as this material is produced only through their exposed surfaces. Dissemination is usually by short trichomes or sections of filament that separate from the plant mass. Short segments of trichomes are referred to as hormogonia.

In the older literature, the filamentous bluegreens have often been placed in several orders that are separated on what appear to be trivial or minor characters.

Tilden (1910) and Drouet (1951) recognized five families of filamentous bluegreens. More recently, Drouet (1968, 1973, 1978) divided the group into three families: the Oscillatoriaceae, without heterocysts or spores; the Nostocaceae, with heterocysts and spores but without true branching; and the Stigonemataceae, with heterocysts and spores and with true branching. His classification is followed here.

While Drouet's *Revision of the Classification of the Oscillatoriaceae* (1968), *Revision of the Nostocaceae with Cylindrical Trichomes* (1973), and *Revision of the Nostocaceae with Constricted Trichomes* (1978) have been followed, his final monograph revising the Stigonemataceae has not yet been completed. Consequently, treatment of the genera *Brachytrichia* and *Mastigocoleus* (Stigonemataceae) are still based upon the older literature (Tilden, 1910).

The following key to the genera and to some species of the Hormogonales seemed more practical than a key to the families followed by
separate keys to the genera. It is an artificial key, dealing only with
genera known to occur or expected to occur in Virginia coastal waters. It
has been constructed solely as an aid to identification and should not be
construed to define the genera included *per se*.

Key to the Genus or Species of the Filamentous Bluegreen Algae

This key leads to the full name of the plant if the genus is represented
by only one species in the Virginia marine flora. Otherwise, it leads to
the genus, and a key to the species will be found following the genus
description. Determination of isolated trichomes or filaments, or of only
a few, may not be possible, as they may not exhibit all the characters es-
sential to recognition of the genus or species.

1. Heterocysts absent 2

1. Heterocysts present 8

2. Trichomes spiral, without cross walls . . . .
. . . . *Spirulina subsalsa* (p. 36)

2. Trichomes with cross walls, usually not spiral ; 3

3. Strictly planktonic, the filaments in fascicles or bundles of various
shapes . . . . *Oscillatoria erythraea* (p. 42)

3. Not strictly planktonic 4

4. With a layer of many granules on each side of each cross
wall 5

4. No granules on cross walls, or at most only two 6

5. End wall of terminal cell thin, never becoming thick. . . . .
. . . . *Arthrospira neapolitana* (p. 45)

5. End wall of terminal cell becoming thickened with age . . .
. . . . *Microcoleus* (p. 46)

6. End wall of terminal cell becoming thickened with age . . . .
. . . . *Oscillatoria* (p. 42)

6. End wall of terminal cell always thin 7

7. Attenuation at tip involving only the terminal cell . . . .
. . . . *Schizothrix* (p. 37)

7. Attenuation at tip involving several to many cells . . . . . .
. . . . *Porphyrosiphon* (p. 40)

8. Plants penetrating shells or limestone; with true branching . .
. . . . *Mastigocoleus testarum* (p. 55)

8. Plants not penetrating limestone 9

9. Trichomes with little or no attenuation at tip 10

9. Trichomes tapering at tip to a slender filament or hair . . . 12

10. Trichomes much constricted at cross walls,
often beadlike 11

10. Trichomes not constricted at cross walls, or only slightly constricted . . . . *Scytonema hofmannii* (p. 53)

11. Cells disk-shaped, shorter than their diameter. . . . . . .
. . . . *Nostoc spumigena* (p. 50)

11. Cells equal to or longer than their diameter . . . .
. . . . *Anabaina oscillarioides* (p. 48)

12. Trichomes with true branching, often with V-shaped angles, always embedded in a common gelatinous matrix . . . . . .
. . . . *Brachytrichia quoyi* (p. 54)

12. Trichomes without true branching, not forming V-shaped angles;
heterocysts basal and sometimes intercalary also; usually a sheath
around each trichome, but sometimes the cluster of trichomes
embedded in a common gelatinous sheath . . . .
. . . . *Calothrix* (p. 50)

FAMILY OSCILLATORIACEAE

Filamentous bluegreen algae without heterocysts and unbranched;
trichomes from 0.4–90 μm in diameter, capable of indefinite growth in
length, divided into cells (except for *Spirulina*) by cross walls that
originate at the periphery and grow inward. Apparently all species
secrete a mucopolysaccharide that may diffuse into the environment or
may form a more or less distinct sheath around the trichomes or fascicle
of trichomes. The nature of the sheath varies with the environment.
Drouet (1968) has concluded that only certain characters of the protoplast are reliable for distinguishing genera and species, and that there

may be no sheath, a soft diffluent one, or a distinct one in virtually any species.

GENUS *SPIRULINA*

Trichomes forming a loose or tight spiral, or merely torulose, without a visible sheath or cross walls, 0.4–4.0 μm in diameter.

SPIRULINA SUBSALSA OERSTED
(Fig. 14)

Plant mass often intense blue-green, the trichomes usually 1–2 μm in diameter in the marine environment, forming tight spirals, especially in

Fig. 14. *Spirulina subsalsa:* two filaments illustrating different degrees of coiling

water of high salinity; of contiguous turns or in part more loosely spiraled, the coils usually 3–5 μm in diameter.

Common in quiet waters and protected areas, intertidal and below, and occasionally in exposed places such as the side of a sea buoy. Often living in the surface layer of intertidal sand in gently sloping beaches protected from wave action. Present along the entire coastline of Virginia from full seawater to fresh water. Of worldwide distribution.

Tilden, 1910, p. 89, pl. 4, fig. 49; Strickland, 1940, p. 633; Humm and Caylor, 1957, p. 235, pl. 1, fig. 11; Zaneveld, 1966, p. 124; Cocke, 1967, p. 33, figs. 67–72.

GENUS *SCHIZOTHRIX*

Schizothrix (*sensu* Drouet, 1968) produces straight, curved, sometimes spiraled trichomes in which tapering at the tip, if any, is usually restricted to the terminal cell. Terminal cell hemispherical to conical, the outer wall not thickened. Sheath firm and distinct or present as a soft mass or apparently absent. Cross walls not granulated or with only one or two distinct granules.

Key to the Species of the Genus *Schizothrix*

1. Terminal cell rotund, not attenuate or only slightly so 2

1. Terminal cell becoming conical 3

2. Trichomes 3.5 μm in diameter or less . . . . . . *S. calcicola*

2. Trichomes 4–60 μm in diameter . . . . . . *S. mexicana*

3. Terminal cell blunt to acute-conical . . . . . . *S. arenaria*

3. Terminal cell long-acuminate . . . . *S. tenerrima*

SCHIZOTHRIX CALCICOLA (AGARDH) GOMONT
(Fig. 15)

Trichomes 0.2–3.5 μm in diameter, the cells 0.2–6.0 μm long. Terminal cell at first cylindrical, becoming rotund, and without a thickened mem-

Fig. 15. *Schizothrix calciola:* at left, a filament enwrapping a branch of *Ectocarpus;* right, a filament with false branches from within the calcium carbonate of a mollusc shell

brane. Trichomes with or without a sheath; if a sheath is present, there may be one to many trichomes within it, or the colony may form a firm, gelatinous hemisphere several centimeters in diameter. Filaments often penetrate shells and other forms of limestone.

In Virginia, *S. calcicola* is known in forms to which the following names were previously applied: *Plectonema terebrans* Bornet and Flahault, penetrating shells and limestone; *P. nostocorum* Bornet and *P. calothrichoides* Gomont as soft, gelatinous masses epiphytic on larger algae or on solid substrata; *Phormidium persicinum* (Reinke) Gomont, forming a thin coating on shells or submerged bottles or in laboratory marine aquaria; and approaching the firm, gelatinous masses of *Phormidium crosbyanum* Tilden when growing in a wooden tank of seawater on the pier of the Virginia Institute of Marine Science. Of worldwide distribution in both salt and fresh water and in hot springs.

C. Agardh, 1810–12, p. 37 (as *Oscillatoria calcicola*); Gomont, 1892, p. 307, pl. 8, figs. 1–3; Tilden, 1910, p. 139, pl. 4, figs. 1–4 (as *Hypheothrix calcicola*); Strickland, 1940, p. 628; Humm, 1963*b*, p. 517 (as the ecophene *Phormidium crosbyanum*); Humm, 1964, p. 309; Zaneveld, 1966, p. 118; Baker and Bold, 1970, p. 21, figs. 2–48, 105–6, 111–12, 114, 117, 119–20.

SCHIZOTHRIX MEXICANA GOMONT
(**Fig. 16**)

Trichomes 4–65 μm in diameter, variable in color, not attenuated at the tips; the terminal cell more or less hemispherical and with a thin outer

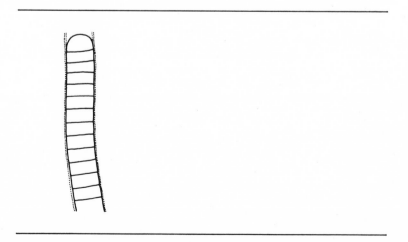

Fig. 16. *Schizothrix mexicana*

membrane. Cells shorter or longer than they are broad, 2–10 μm long; the cross walls not granulated.

In Virginia this species has been found as its ecophene *Lyngbya gracilis* Meneghini, growing around the base of the salt marsh grass *Spartina* near Gloucester Point. Of worldwide distribution in tropical and temperate regions in both salt and fresh water.

Gomont, 1892, p. 124, pl. 2, fig. 20; Tilden, 1910, p. 117, pl. 5, fig. 36 (as *Lyngbya gracilis*); Drouet, 1968, p. 87, figs. 20–22.

SCHIZOTHRIX ARENARIA (BERKELEY) GOMONT
(**Fig. 17**)

Trichomes 1–6 μm in diameter, the cells isodiametric or longer than wide, 2–10 μm long, the tip cells distinctly conical and with a thin

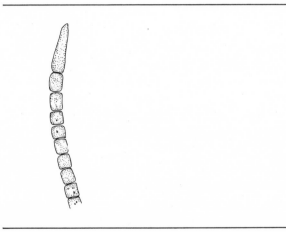

Fig. 17. *Schizothrix arenaria*

membrane. Cross walls slightly to distinctly constricted, without granules or sometimes with a single granule against each wall.

In Virginia, the most common form produces trichomes that are 2.5–6.0 μm in diameter, with many trichomes in the sheath. The plants are mat forming and often mixed with other bluegreens, especially *Schizothrix tenerrima*. They are abundant on oyster reefs and wet muddy sand throughout the Virginia coastal areas. The old name for this form is *Microcoleus chthonoplastes* (Mertens) Zanardini.

Of worldwide distribution.

Gomont, 1892, p. 353, pl. 14, figs. 5–8; Tilden, 1910, p. 155, pl. 6, fig. 28; Fremy, 1934, p. 67, pl. 17, fig. 7 (all as *M. chthonoplastes*);

Drouet, 1968, p. 109, figs. 28–34; Baker and Bold, 1970, p. 30, figs. 49–53.

SCHIZOTHRIX TENERRIMA (GOMONT) DROUET
(Fig. 18)

Trichomes 1–6 μm in diameter, variable in color, usually constricted at the nodes, the cells 3–12 μm long, the cross walls not granulated.

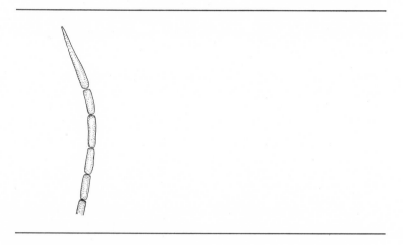

Fig. 18.   *Schizothrix tenerrima*

Terminal cells cylindrical at first but becoming long-conical with a hair-like tip, the outer membrane thin.

In Virginia the most common form of the plant has trichomes 1.5–2.0 μm in diameter, the cells 2.5–6.0 μm long, and it forms extensive mats on intertidal muddy sand or on shells, submerged wood, or pilings throughout the Chesapeake Bay and along the Eastern Shore. The plants are more abundant in brackish water than in full seawater. It is widely distributed in temperate and tropical regions.

Gomont, 1892, p. 355, pl. 14, figs. 9–11; Tilden, 1910, p. 155, pl. 6, fig. 27; Humm and Caylor, 1957, p. 238, pl. 2, fig. 9; Zaneveld, 1966, p. 117 (all as *Microcoleus tenerrimus* Gomont); Drouet, 1968, p. 135, figs. 42–45.

GENUS PORPHYROSIPHON

Trichomes attenuated in mature tips through several cells, the terminal cell hemispherical to conical, its outer wall not thickened. Cross walls not granulated.

PORPHYROSIPHON NOTARISII (MENEGHINI) KÜTZING

(Fig. 19)

Trichomes 3–40 μm in diameter, the cells shorter or longer than broad, 3–5 μm long, the cross walls not granulated. Mature tips attenuated through several to many cells; the tip cell at first hemispherical, becoming conical, its outer membrane not thickened.

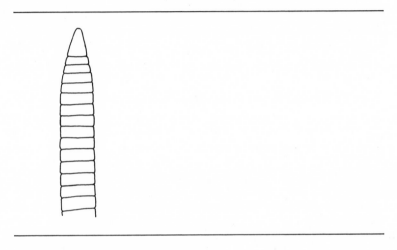

Fig. 19.   *Porphyrosiphon notarisii*

In Virginia marine and brackish waters this species occurs as two forms previously known as *Oscillatoria nigro-viridis* Thwaites and *O. subuliformis* Thwaites. The former is the more abundant of the two, but both inhabit salt marshes, pilings, oyster shells, and other objects in the intertidal zone or below throughout the Chesapeake Bay and along the Eastern Shore. The *O. nigro-viridis* form of *P. notarisii* produces trichomes 7–11 μm in diameter. *O. subuliformis* trichomes are generally 4.5–6.5 μm in diameter.

Of worldwide distribution in salt and fresh water, in hot springs, and on soil, rocks, and other surfaces. It will withstand indefinite periods of dessication.

O. subuliformis: Gomont, 1892, p. 226, pl. 7, fig. 10; Tilden, 1910, p. 77, pl. 4, fig. 27. *O. nigro-viridis:* Gomont, 1892, p. 217, pl. 6, fig. 20; Tilden, 1910, p. 69, pl. 4, fig. 12; Humm and Caylor, 1957, p. 234, pl. 1, fig. 8; Cocke, 1967, p. 37, fig. 73. *Porphyrosiphon notarisii:* Drouet, 1968, p. 143, figs. 65–75; Baker and Bold, 1970, p. 40, figs. 89–94.

Porphyrosiphon kurzii (Zeller) Drouet is known from Florida to

Beaufort, North Carolina, and may occur along the Eastern Shore of Virginia, at least during the summer. It is a strictly marine species.

Porphyrosiphon splendidus (Greville) Drouet (*Oscillatoria splendida* Greville) is known from several freshwater habitats in Virginia, but since it appears to be a strictly freshwater species, it is not treated here.

GENUS *OSCILLATORIA*

Trichomes in which the mature terminal cells produce a distinctly thickened outer wall. The cross walls are not granulated, or they show only one or two granules (*sensu* Drouet, 1968).

Key to the Species of the Genus *Oscillatoria*

1. Strictly planktonic, the filaments in fasciculate bundles; the cells often with pseudovacuoles . . . . *O. erythraea*

1. Not strictly planktonic; without pseudovacuoles 2

2. Cells mostly isodiametric or a little longer; thickened outer wall of end cell hemispherical or slightly conical . . . .
    . . . . *O. submembranacea*

2. Cells mostly shorter than broad; end of trichome only slightly tapered or not tapered . . . . *O. lutea*

OSCILLATORIA ERYTHRAEA (EHRENBERG) KÜTZING

(Fig. 20)

Trichomes 3–30 μm in diameter, the cells 2–27 μm long, often with large pseudovacuoles, the cross walls not granulated. Terminal cell cylindrical, swollen, or truncate-conical, developing a thickened outer membrane with age. Trichomes typically forming colonies several millimeters in length or diameter, and normally restricted to the open-sea plankton.

This species often forms extensive plankton blooms in warm or tropical waters, especially in areas where nitrogen compounds are in low concentration, as it is known to be able to fix nitrogen. It often discolors the water amber or reddish in great patches, sometimes many miles in diameter, and its macroscopically visible scalelike colonies are often re-

Fig. 20. *Oscillatoria erythraea*

ferred to as "sea sawdust." It is responsible for the color and name of the Red Sea, and its blooms may cause fish kill when the cells lyse.

Colonies are sometimes found adhering temporarily to pilings, rocks, or other algae or trapped in beach sand, but the species apparently does not survive in these situations. It appears to be difficult to grow in culture.

Occasional in the plankton of offshore Virginia waters and in the lower Chesapeake Bay during late summer, probably derived from the Gulf Stream.

This species has been referred to several species, based upon minor differences, in the genera *Trichodesmium* and *Skujaella*.

Gomont, 1892, p. 217, pl. 6, figs. 2–4; Tilden, 1910, p. 84, pl. 4, figs. 41–42 (both as *Trichodesmium thiebautii* Gomont); Drouet, 1968, p. 212, figs. 130–31.

OSCILLATORIA SUBMEMBRANACEA ARDISSONE & STRAFFORELLO

(Fig. 21)

Trichomes 2.5–9.0 μm in diameter, the cells 3–11 μm long, the terminal cell truncate-cylindrical or truncate-conical, the outer membrane at first thin but becoming thick, the tips of the trichomes not attenuate or only slightly so.

In Virginia this species has been recorded in brackish and salt water as its ecophene *Symploca atlantica* Gomont, forming a blackish-green

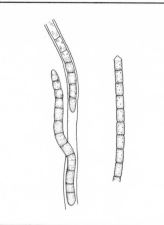

Fig. 21. *Oscillatoria submembranacea,* trichomes showing false branching and two types of terminal cells

coating on barnacles from pilings of the pier at the Virginia Institute of Marine Science, Gloucester Point.

Widely distributed in tropical and temperate regions in fresh and salt water and on damp soil. It survives long periods of dessication.

Gomont, 1892, p. 109, pl. 2, fig. 6; Tilden, 1910, p. 129, pl. 5, fig. 48 (both as *Symploca atlantica*) ; Drouet, 1968, p. 203, figs. 62–64.

OSCILLATORIA LUTEA C. AGARDH

Trichomes 2.5–10 μm in diameter; the cells 1–7 μm long, usually shorter than wide. Terminal cell broadly truncate-conical, the outer wall at first thin but becoming thickened. Cross walls not granulated, without constrictions at the nodes or only slightly and irregularly constricted.

In Virginia this species has been found on the leaves of eel grass *Zostera marina* L. near Gloucester Point and around the base of larger algae at Fort Wool. These collections, having a distinct sheath, agreed with the synonym *Lyngbya lutea* (C. Agardh) Gomont.

Widely distributed in tropical and temperate regions in both salt and fresh water and on soil and other wet substrata.

Gomont, 1892, p. 141, pl. 3, figs. 12–13; Tilden, 1910, p. 114, pl. 5, figs. 30–31; Humm and Caylor, 1957, p. 235, pl. 2, fig. 2 (all three as *Lyngbya lutea*); Drouet, 1968, p. 185, figs. 51–57; Baker and Bold, 1970, figs. 99–104, 107–110, 116, 118, 121–22.

GENUS *ARTHROSPIRA*

Trichomes straight, variously curved or spiral, often attenuated at the tip, the outer wall of the terminal cell not thickened. Cross walls densely lined with granules.

ARTHROSPIRA NEAPOLITANA (KÜTZING) DROUET

(Fig. 22)

Trichomes 2–10 μm in diameter; the cells about one-third as long as they are wide, 1.5–4.0 μm long. Terminal cell conical, up to six times

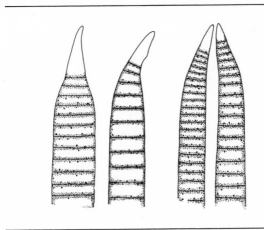

Fig. 22. *Arthrospira neapolitana*

as long as it is broad; the outer wall not thick. Constrictions may occur at the cross walls.

Known from Virginia through a culture obtained from the York River at Yorktown by Dr. J. C. Strickland, and from several fresh-water habitats in the Commonwealth.

Widely distributed in fresh and brackish waters in temperate and tropical regions and on persistently moist soil high in organic matter.

Gomont, 1892, p. 229, pl. 7, figs. 14–15; Tilden, 1910, p. 79, pl. 4, fig. 32 (both as *Oscillatoria brevis* Kützing); Drouet, 1968, p. 219, figs. 86–88 (as *Arthrospira brevis* [Kützing] Drouet).

GENUS *MICROCOLEUS*

Trichomes with a layer of numerous granules on each side of the cross walls. Tips of trichomes often attenuated, the end cell truncate, hemispherical, or depressed-conical; outer wall of the tip cell thickened with age.

Key to the Species of the Genus *Microcoleus*

Dense protoplasm and granules along both side and cross walls   .   .   .
.   .   .   . *M. lyngbyaceus*
Dense protoplasm and granules along cross walls only   .   .   .   .   .
.   .   .   . *M. vaginatus*

MICROCOLEUS LYNGBYACEUS (KÜTZING) CROUAN

(**Fig. 23**)

Trichomes 3.5–80 μm in diameter. Cells usually shorter than they are broad, 1.5–8 μm long, with constrictions at the cross walls. Tips of trichomes may or may not be attenuated. Terminal cells with rounded ends and thin-walled at first, but becoming thick-walled and sometimes conical with age.

   M. lyngbyaceus is one of the most common bluegreens in the marine- and brackish-water areas of the Chesapeake Bay and the Eastern Shore.

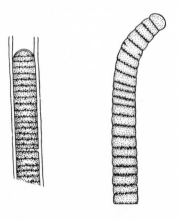

Fig. 23.   *Microcoleus lyngbyaceus,* trichomes with and without a sheath

At least six forms of this species that have been given different names in the past are known locally: *Oscillatoria corallinae* Gomont has been found on the leaves of *Zostera* in the lower Chesapeake Bay and on pilings in the Norfolk area. *Lyngbya semiplena* (C. Agardh) J. Agardh is widely distributed in salt marshes, on oyster shells, and on the inter-tidal zone of seawalls. *Lyngbya aestuarii* (Mertens) Lyngbye is com-mon on mud in salt marshes. *Lyngbya confervoides* C. Agardh is abundant in the intertidal zone and often in a pure stand on creosoted pilings. *Hydrocoleum lyngbyaceum* Kützing was found under a stand of *Gelidium crinale* on conglomerate rocks and oyster shells in the York River at Yorktown. *Hydrocoleum glutinosum* (C. Agardh) Gomont produced a thin *Phormidium*-like layer on the concave side of a piece of broken bottle at Wachapreague on the Eastern Shore. These ecophenes are described and illustrated in Tilden (1910).

Drouet, 1968, p. 262, figs. 101–29; Baker and Bold, 1970, figs. 81–86, 125.

MICROCOLEUS VAGINATUS (VAUCHER) GOMONT

Trichomes 2.5–9.0 μm in diameter, the tips usually attenuated through several cells. Cells isodiametric, shorter or longer than broad, 1–10 μm long, not constricted (or rarely) at the cross walls. Cross walls with a layer of granules on each side. Terminal cell variable in shape, the outer wall becoming thickened with age.

Several forms of this species are known from Virginia. One of them, previously known as *Phormidium uncinatum* (C. Agardh) Gomont, once became abundant in an aquarium at the Virginia Institute of Marine Science in the laboratory of Dr. Marvin Wass. Many collec-tions have been made from fresh water in Virginia.

Of worldwide distribution in fresh water, but also in brackish water. It survives long periods of dessication.

Gomont, 1892, p. 355; Cocke, 1967, p. 75, fig. 166; Drouet, 1968, p. 226, figs. 89–99; Baker and Bold, 1970, figs. 54–80, 113, 115, 123–24.

FAMILY NOSTOCACEAE

Trichomes unbranched, uniseriate, with heterocysts and with or without spores. The presence or absence and the relative abundance of both heterocysts and spores are strongly influenced by the environment. Production of heterocysts is stimulated by low availability of fixed nitrogen and is inhibited by the opposite condition.

Dr. Francis Drouet has revised the Nostocaceae in two monographs. The first (1973) dealt with the species having cylindrical trichomes, that is, trichomes in which the cells are not sharply constricted at the cross walls. He eliminated all those genera that were based upon sheath characteristics and other differences that vary significantly with the environment. He recognized only three genera: *Calothrix, Scytonema,* and *Raphidiopsis.* Only the first two are represented in marine habitats.

The second revision of the Nostocaceae (Drouet, 1978) dealt with the remainder, all those species with constricted trichomes. In this work, he recognized only two genera: *Nostoc* and *Anabaina* (incorrectly spelled *Anabaena* in most publications). Each of these genera is represented by one species that is common in marine habitats.

GENUS *ANABAINA*

Trichomes much constricted at the cross walls, torulose or moniliform, the terminal vegetative cells at first hemispherical or spherical, becoming blunt-conical or acute-conical with age. Heterocysts terminal or intercalary, the latter spherical to cylindrical with rounded ends, the terminal heterocysts usually conical, sometimes spherical or ovoid. Spores produced, one after another, in a series in the trichome; spherical, ovoid, or cylindrical when mature. Extracellular polysaccharides may be diffluent and invisible, confluent into a soft layer or mass that serves as a common gelatinous matrix for a clone of trichomes, or as a firm and distinct sheath around each trichome.

The genus *Anabaina* was established by Bory de Saint-Vincent in 1822. A few years later the name was mispelled *Anabaena* by another author, and subsequent authors have used the incorrect spelling ever since. "Anabaena" is not available for conservation, hence it is necessary to go back to Bory de St.-Vincent's original and correct name.

Two species in the older nomenclature have been reported for Virginia or recorded as herbarium specimens: *A. torulosa* (Carmichael) Lagerheim and *A. inaequalis* (Kützing) Bornet and Flahault.

ANABAINA OSCILLARIOIDES BORY

(Fig. 24)

Trichomes 2–12 μm in diameter, blue-green, yellow-green, olive, brown, red, or violet, torulose, deeply constricted at the cross walls; vege-

Fig. 24. *Anabaina oscillarioides,* a trichome with a heterocyst and two spores

tative cells nearly spherical or longer or shorter than their diameter, 2–10 μm long, the terminal cells at first spherical, becoming conical with maturity. Heterocysts intercalary in origin, spherical to ovoid with rounded ends, 3–14 μm in diameter. Spores produced serially adjacent to or remote from a heterocyst, to 20 μm in diameter when mature.

Anabaina oscillarioides is abundant in shallow-water marine habitats but also occurs in brackish and fresh water. In warm seawater, spores are not produced.

In Virginia this species has been found in abundance in the lower York River as scattered filaments or groups of filaments in intertidal sand or muddy sand, especially in or near salt marshes; it has been found on rocks near the highway bridge at Yorktown. In the Virginia collections the trichomes are mostly 4–5 μm in diameter, the heterocysts about 6 μm in diameter, the sheath material invisible or evident as a confluent layer.

The species is easily found by skimming the surface of intertidal sand or muddy sand to a depth of a few millimeters, shaking the material in a small volume of water, and pouring off the water as soon as the heavy particles have settled. If the water is then allowed to stand, trichomes of *A. oscillarioides* and various other bluegreen algae will settle to the bottom. *Agmenellum* is another genus that will be found by this procedure.

Bory de St.-Vincent 1822, p. 308; Tilden 1910, p. 191, pl. 9, fig. 16 (as *Anabaena inaequalis*); p. 192, pl. 9, fig. 18 (as *A. torulosa*); Desikachary 1959, 417, pl. 71, fig. 7 (*sensu strictu*); Drouet 1978, p. 182, figs. 32–42.

GENUS *NOSTOC*

Trichomes much constricted at the cross walls, torulose or moniliform, the terminal vegetative cells more or less hemispherical or discoid, becoming almost spherical, ovoid, or cylindrical with a rounded end, or discoid with a rounded end. Heterocysts and spores spherical, discoid, ovoid, or cylindrical. Extracellular polysaccharide diffluent, soft and confluent, or forming a distinct sheath.

NOSTOC SPUMIGENA (MERTENS) DROUET

Trichomes 3–20 μm in diameter, much constricted at the cross walls, blue-green, yellow-green, olive, brown, red, or violet. Vegetative cells disciform or compressed-spherical, mostly shorter than their diameter, the terminal cell hemispherical to subspherical. Heterocysts intercalary disciform or compressed-spherical, 3–20 μm in diameter. Spores disciform to spherical, intercalary in origin, produced serially. Extracellular polysaccharides not in evidence, or in a confluent layer or, rarely, in the form of a distinct sheath.

Nostoc spumigena is common in the sea, usually on the bottom or on a solid surface, but sometimes in the plankton, where it often has pseudovacuoles. In Virginia it is common on shells, stones, and other algae along the York River cliffs above the bridge near Yorktown, and in other similar localities throughout the Chesapeake Bay. Prior to the appearance of Drouet's 1978 monograph, this material was referred to *Nodularia harveyana* (Thwaites) Thuret.

Mertens in Jürgens, 1816–22 (1822), folio 15, No. 4; Bornet and Flahault 1888, p. 245; Tilden 1910, p. 182, pl. 9. figs. 1–2 (as *Nodularia harveyana*); Drouet 1978, p. 126, figs. 24–25.

GENUS *CALOTHRIX*

Trichomes tapering at the apex to a hair tip (unless broken off), in some cases tapering from base to apex; unbranched or with false branches, the latter often originating with the cell just below an intercalary heterocyst. Heterocysts are nearly always present, basal only, or with intercalary heterocysts as well. Spores are sometimes present, especially in plants from brackish or fresh water.

Depending upon the degree of development and persistence of the sheath, the plants may be in the form of a small group of erect filaments, in dense tufts, or a turf. Each trichome may have a separate sheath, or false branches may remain enclosed in the parent sheath in their lower parts, producing fasciculate groups of filaments with the upper parts free and enclosed only in the individual sheath (the old genus *Dichothrix*); or the trichomes may have basal heterocysts only, with masses of trichomes radiating from the center of a spherical or hemispherical common gelatinous matrix that may become hollow with age (the old genus *Rivularia*).

Calothrix grows in the intertidal zone, where it often forms a high black band on seawalls, pilings, or other woodwork, a band that may be mistaken as the high-water mark of an old oil spill. On intertidal rocks or breakwaters of the open sea, it is often found in the *Dichothrix* or *Rivularia* forms. *Calothrix* also thrives below low tide, perhaps to the maximum depths at which any benthic algae are able to grow. It can be found as an epiphyte on nearly all specimens of larger algae. On the pelagic species of *Sargassum* in the Sargasso Sea, it produces the condition known as "tar spot," a name suggested by the black *Dichothrix*-form of epiphytic tufts. It is common on invertebrate animals, rocks, and other solid substrata, and it sometimes penetrates limestone.

Drouet (1973) recognizes only two species, both of which occur in brackish water but one of which is characteristic of fresh water and the other of seawater.

Key to the Species of the Genus *Calothrix*

Trichomes attenuated gradually to the slender tip; in fresh and brackish water . . . . *C. parietina*

Trichomes abruptly attenuated near the tip; in brackish water and seawater . . . . *C. crustacea*

CALOTHRIX PARIETINA (NÄGELI) THURET

Trichomes usually blue-green, yellow-green, or olive green, but they may be red, violet, or brownish; 3–24 μm in diameter, long-tapering to a hair tip, usually with a basal heterocyst and often with intercalary heterocysts. The trichomes may be constricted or not constricted at the cross walls; naked or with a sheath around each trichome, the sheaths often branched, or many trichomes embedded in a common gelatinous matrix that results from coalesced sheaths.

Common in the low-salinity areas of the Chesapeake Bay and up the river systems into fresh water. On rocks, wood, unconsolidated sediments; epiphytic, and upon animals. Of worldwide distribution.

Thuret, 1875, p. 381; Bornet and Flahault 1886–88, p. 366; Tilden, 1910, p. 269, pl. 18, fig. 12; Cocke, 1967, p. 160, fig. 310.

CALOTHRIX CRUSTACEA SCHOUSBOE AND THURET

(Fig. 25)

Trichomes usually blue-green or olive green but they may be other colors, 3–22 μm in diameter, tapering at the upper ends to a hair tip (if

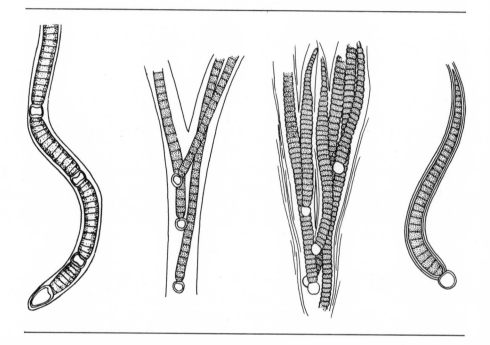

Fig. 25. *Calothrix crustacea:* the trichome at the left with both basal and intercalary heterocysts; the two groups in the center showing branching of the sheath (false branching); the trichome at the right with a basal heterocyst only and the characteristically tapered tip

tip is intact). Basal heterocysts are usually present and often intercalary heterocysts also. Constrictions may or may not be present at the cross walls. Trichomes may be without a sheath, there may be a discrete sheath around each, or groups or masses of trichomes may be embedded in a common gelatinous matrix.

  C. crustacea is one of the most abundant species of bluegreens in the lower Chesapeake Bay and along the entire seacoast of Virginia on all kinds of submerged solid surfaces, as an epiphyte, and on animals. It is abundant intertidally where it forms a black zone. In salt marshes it may produce an upper band of epiphytes on the stems of the salt marsh plants *Spartina* and *Juncus*. It is usually present on the pelagic *Sargassum* plants that drift into the Chesapeake Bay during the summer or wash ashore along the outer beaches in the form known as "tar spot." Along the Eastern Shore on exposed intertidal rocks or concrete, it may form small, gelatinous spheres, formerly referred to the genus *Rivularia*. Its distribution is worldwide.

Bornet and Flahault, 1886–88, p. 359; Tilden, 1910, p. 264, pl. 17, figs. 2–6; Desikachary, 1959, p. 523, pl. 111, figs. 10–11; Cocke, 1967, p. 156, fig. 304.

GENUS *SCYTONEMA*

Trichomes usually blue-green, olive green, or yellow-green, but they may be brownish, red, or violet; usually cylindrical, but there may be some constriction at the cross walls; 3–30 μm in diameter, sometimes attenuated at the tip through several cells; the cross walls not granulated. Cells longer or shorter than their diameters, 3–20 μm long. Heterocysts terminal or intercalary, usually cylindrical, 4–30 μm in diameter. Spores cylindrical, seriate, the walls becoming yellow to brown. Sheaths usually distinct, often branched, usually with a single trichome.

Drouet (1973) recognizes only one species.

SCYTONEMA HOFMANNII C. AGARDH

On wooden braces in the upper intertidal zone under the fishing pier at Ocean View, Norfolk, in the form of blackish-green tufts or as a turf. This apparently is the only record in seawater in Virginia but it is widely distributed in fresh water in the state and in damp places on land. Since it grows equally well in fresh water and the sea, it probably occurs throughout the Chesapeake Bay and along the Eastern Shore. It is of worldwide distribution.

Drouet, 1973, p. 63, figs. 1–27.

FAMILY STIGONEMATACEAE

Trichomes with true branching, uniseriate or multiseriate, with heterocysts and sometimes spores. Morphologically, this is the most highly evolved family of the bluegreens.

GENUS *BRACHYTRICHIA*

Plants producing gelatinous masses, spheres, or hemispheres in which the trichomes are embedded, at first solid, but becoming hollow and more irregular in shape. Trichomes parallel in the sheath material, much branched, the tips tapering to a hair. Heterocysts intercalary and irregular in arrangement.

BRACHYTRICHIA QUOYI BORNET AND FLAHAULT

(Fig. 26)

Plants forming gelatinous masses from a few millimeters to several centimeters in diameter; often epiphytic, intertidal and below, the larger

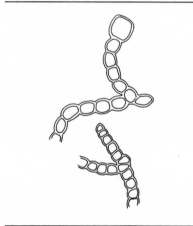

Fig. 26. *Brachytrichia quoyi,* segments of trichomes showing branching and a terminal heterocyst

colonies folded and irregular at the surface, hollow. Filaments 25–55 μm in diameter; cells near the ends tending to be thick and discoidal, the inner cells more oval in shape. Heterocysts of a little greater diameter than the trichomes.

Known from the south side of Cape Cod (summer only) to the West Indies. Though apparently not yet recorded for Virginia, it probably occurs along the Eastern Shore. It has been found in North Carolina, South Carolina, and Florida, where it usually occurs on structures exposed to the open sea, high salinity, and surf.

Bornet and Flahault, 1886–88, p. 373; Tilden, 1910, p. 294, pl. 20, fig. 18; Aziz and Humm, 1962, p. 58; Zaneveld, 1966, p. 114; Cocke, 1967, p. 172, fig. 326; Blackwelder, 1975, p. 270.

GENUS *MASTIGOCOLEUS*

Plants penetrating shells and other forms of calcium carbonate; branched, the branches uniseriate, of two kinds: one cylindrical and the other flagelliform and tapering to a hairlike tip. Heterocysts terminal on short, cylindrical branches. There is but one species.

MASTIGOCOLEUS TESTARUM LAGERHEIM

(Fig. 27)

Filaments curved, branched, 6–10 µm in diameter, the branches of two kinds: short (of one to four cells), terminating in a heterocyst, and longer branches ending in an extended hairlike process. Trichomes uniseriate, 3.5–6.0 µm in diameter; heterocysts 9–12 µm in diameter.

Common in old mollusc shells throughout the Chesapeake Bay and along the Eastern Shore. Widely distributed.

Lagerheim, 1886, p. 65, pl. 1; Tilden, 1910, p. 237, pl. 14, fig. 12; Aziz and Humm, 1962, p. 58; Cocke, 1967, p. 146, fig. 291.

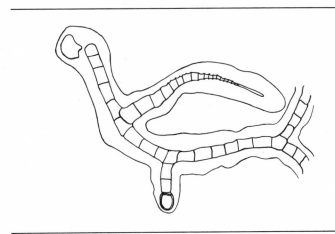

Fig. 27. *Mastigocoleus testarum*, part of a plant from a decalcified mollusc shell showing a terminal heterocyst and a flagelliform branch

RHODOPHYTA

The red algae, Rhodophyta, range in color from various shades of red to olive green, purplish-green, and yellow-brown because the red pigment for which they are noted, phycoerythrin, is often masked by their chlorophylls (*a* and *c*) or by various carotenoid pigments, partly because of hereditary influences and partly because of environment. In general, the red pigment is best developed under low light intensity, but some species such as *Hypnea musciformis* and *Gracilaria foliifera* are almost always predominantly green, usually olive green or purplish-green. The majority, however, are plainly red.

The red algae are nearly all multicellular plants and are fundamentally filamentous in structure, although many become pseudoparenchymatous. The reproductive cells are nonmotile (except perhaps for amoeboid motility of the spermatia and spores), and all are oogamous, the egg cell never leaving the female plant. Nearly all are marine (only about 3% occur in fresh water) and they occur from the intertidal zone to a depth of 200 meters in the clearest sea water. Some precipitate calcium carbonate in the cell walls (Corallinaceae) and contribute significantly to calcareous sediments and reefs.

Food is stored as floridean starch, a type found only in the red algae, and also as floridoside, a galactoside of glycerol. The outer part of the cell walls is made up of a thick layer of polysaccharide, agar, carrageenan, or another type of galactan. Agar and most types of carrageenan form thermally reversible gels. They can be extracted by drying the plants and boiling them in water or treating them under steam pressure. Some cell-wall polysaccharides of the red algae are not soluble in boiling water and do not form thermally reversible gels (Humm, 1951a, 1951b).

Though often regarded as characteristic of tropical waters, the red algae are widely distributed in the oceans of the world. Setchell (1915) pointed out that only 22% of the known species were recorded from the tropics; 34% were native to temperate or cold waters of the northern hemisphere, and 44% were native to temperate and cold waters of the southern hemisphere.

The red algae are a well-defined group, so that, traditionally, all are placed in the same class, the Rhodophyceae.

Key to the Genus or Species of the Rhodophyta

1. Plants completely calcified 2

1. Plants not calcified 6

2. Forming a thin disk on seagrass leaves or on other algae 3

2. Growing erect 5

3. Tetrasporangial conceptacles with few to many pores . . . . *Melobesia membranacea* (p. 80)

3. Tetrasporangial conceptacles with only one pore 4

4. A trichocyte terminating cell rows on surface of crust . . . . *Fosliella farinosa* (p. 81)

4. Without trichocytes . . . . *Fosliella lejolisii* (p. 81)

5. Ultimate branching pinnate . . *Corallina officinalis* (p. 83)

5. Ultimate branching dichotomous . . *Jania capillacea* (p. 83)

6. Plants as microscopic disks on larger algae and seagrass leaves . . . . *Erythrocladia subintegra* (p. 63)

6. Plants growing erect 7

7. Plants filamentous, unbranched 8

7. Plants not filamentous; or if filamentous, branched 9

8. Epiphytic, uniseriate . . . . *Erythrotrichia* (p. 61)

8. Not epiphytic; pluriseriate in upper parts; intertidal . . . . *Bangia fuscopurpurea* (p. 66)

9. Plants in the form of a flat sheet 10

9. Plants not a flat sheet 12

10. Blades with midrib, at least near the base 11

10. Blades without a midrib, thin, strictly intertidal . . . . . .
. . . . *Porphyra* (p. 66)

11. Intertidal; olive green or purplish-green; blades less than 1 cm
long . . . . *Caloglossa leprieurii* (p. 109)

11. Below low tide; blades rose-red, 15–30 cm long or more . . .
. . . . *Grinnellia americana* (p. 110)

12. Plants entirely uniseriate 13

12. Not entirely uniseriate; or if uniseriate, with corticating cells at
least at the nodes 21

13. Microscopic and epiphytic; less than 5 mm tall 14

13. Not microscopic; over 5 mm tall 16

14. Without a slender cytoplasmic connection between cells; walls relatively thick, gelatinous 15

14. With a slender cytoplasmic connection between cells; the walls
relatively thin . . . . *Acrochaetium* (p. 71)

15. Cells ovate-elongate, greenish-gray . . . .
. . . . *Asterocytis ramosa* (p. 63)

15. Cells red, shorter than the diameter or isodiametric, squarish
. . . . *Goniotrichum alsidii* (p. 64)

16. Uniseriate throughout, uncorticated 17

16. Uniseriate and corticated, or polysiphonous, or appearing to be
parenchymatous 21

17. Filaments of barrel-shaped cells, 120–300 μm in diameter . .
. . . . *Griffithsia tenuis* (p. 102)

17. Filaments of cylindrical cells 18

18. Gland cells present 19

18. Without gland cells 20

19. Upper branches opposite, gland cells occasional on the branchlets
. . . . *Antithamnion cruciatum* (p. 97)

19. A gland cell present at every (or nearly every) node . . . . .
. . . . *Trailliella intricata* (p. 96)

20. Branchlets mostly opposite and pinnate on the erect branches .
. . . . *Spermothamnion turneri* (p. 101)

20. Branching alternate or dichotomous . . . .
. . . . *Callithamnion* (p. 99)

21. Uniseriate, corticated at nodes only or corticating cells covering
entire axis . . . . *Ceramium* (p. 102)

21. Not uniseriate throughout 22

22. Ultimate branchlets or branch tips uniseriate 23

22. No uniseriate branchlets or branch tips (except trichoblasts)
. . . . 25

23. Strictly intertidal; olive green; branches ending in uniseriate tips
. . . . *Bostrichia rivularis* (p. 119)

23. Not intertidal; only the ultimate branchlets uniseriate . . . 24

24. Branches covered with abundant fine red uniseriate filaments . .
. . . . *Dasya baillouviana* (p. 113)

24. Ultimate branchlets uniseriate, but corticated at nodes like *Ceramium* . . . . *Spyridia filamentosa* (p. 107)

25. Axes hollow at the center 26

25. Axes cellular throughout, not hollow 28

26. Axes of barrel-shaped segments with cellular cross walls at the
constrictions . . . . *Champia parvula* (p. 94)

26. Axes not consisting of barrel-shaped segments 27

27. Hollow center with many loose, colorless, anastomosing filaments
. . . . *Solieria tenera* (p. 84)

27. Center without filaments . . *Lomentaria baileyana* (p. 93)

28. Uncorticated, but with four to sixteen pericentral cells; branch tips
usually with trichoblasts . . . . . *Polysiphonia* (p. 115)

28. Corticated, or apparently parenchymatous 29

29. Branch tips with a cluster of trichoblasts . . . .
     . . . . *Chondria* (p. 121)

29. Branch tips bare 30

30. Plants blackish, wiry; axes 0.5 mm diameter or less . . . 31

30. Plants greenish to red; main axes over 1 mm in diameter . . .
     32

31. Branching partly or mainly dichotomous . . . .
     . . . . *Gymnogongrus griffithsiae* (p. 91)

31. Branching pinnate or irregular . . . . *Gelidium* (p. 78)

32. Spinelike branchlets numerous, the main branches often ending in
     a hooked tip . . . . *Hypnea musciformis* (p. 86)

32. Without spinelike branchlets or hooked tips . . . .
     . . . . *Gracilaria* (p. 88)

SUBCLASS BANGIOPHYCIDAE

The subclass Bangiophycidae is a small, primitive group of red algae that are relatively simple in terms of morphology, reproduction, life history, and even cytology. Some of them have so many characteristics in common with bluegreen algae that they were included in that group in the older literature. Tilden (1910, pp. 295–97), for example, lists *Goniotrichum elegans* (= *G. alsidii*), *Asterocytis ramosa,* and *Porphyridium cruentum* under "families and genera not well understood" in her well-known work on the Myxophyceae of North America.

Apparently there is no sexual reproduction in the Bangiophycidae. The cells usually have but one set of chromosomes, a single nucleus, and one stellate or parietal chromatophore. Growth (cell division) is intercalary or diffuse, as there is no differentiation of a meristematic region. There are no cytoplasmic connections between cells. Asexual reproduction is by monospores formed in monosporangia or by neutral spores not produced in sporangia.

The plants are single cells, unbranched or branched filaments, a filiform cylinder of cells, a disk, or sheets of one or two cell layers. In the order Bangiales the plants occur in two morphological types, and reproduction is controlled by season, principally water temperature, but also photoperiod.

ORDER GONIOTRICHALES

Plants small, branched or unbranched, epiphytic, endophytic, or on animals, sometimes on nonliving substrata. The cells have a single stellate chromatophore with a central pyrenoid. Reproduction is by transformation of the protoplast of a vegetative cell into a single, neutral spore that escapes from the old cell wall.

FAMILY GONIOTRICHACEAE

Plants filamentous, unbranched or branched, growing erect, or forming a prostrate disk of closely appressed filaments, usually epiphytic.

GENUS *ERYTHROTRICHIA*

Plants of unbranched filaments, erect, attached by a modified basal cell or a basal disk of a few cells; uniseriate at first, but often producing longitudinal walls later and becoming pluriseriate in part. Vegetative cells producing monospores by oblique, unequal cell division. Sexual reproduction is described in the older literature but is in need of confirmation.

Key to the Species of the Genus *Erythrotrichia*

Base consists of a single cell . . . . *E. carnea*
Base of short, branched filaments . . . . *E. rhizoidea*

ERYTHROTRICHIA CARNEA (DILLWYN) J. AGARDH

(**Fig. 28**)

Plants of unbranched filaments, attached by a lobed basal cell, 1–5 mm tall or more; cells more slender below than above in mature plants, 9–13 μm in diameter and 15–40 μm long in the lower parts, 16–26 μm in diameter and 16–32 μm long in the upper parts. The plants are usually uniseriate, but occasionally a few longitudinal walls develop in the upper parts in large plants. Chromatophores axial, with radiating lobes and a central pyrenoid.

Abundant in Virginia as an epiphyte of larger algae and seagrass

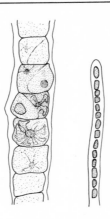

Fig. 28. *Erythrotrichia carnea:* A highly magnified segment of a filament at the left, showing the formation of a monospore and the stellate, radiating chromatophore; the filament at the right is from a young plant.

leaves in the Chesapeake Bay and along the Eastern Shore. Known from the Caribbean Sea to the Canadian maritime provinces and James Bay.

J. Agardh, 1883, p. 15, pl. 1, figs. 8–10; Hoyt, 1917–18, p. 466, fig. 24; Taylor, 1957, p. 202, pl. 28, figs. 13–15; Humm and Caylor, 1957, p. 250, pl. 7, figs. 1–3.

ERYTHROTRICHIA RHIZOIDEA CLELAND

Plants epiphytic, attached at the base by a group of branched, multi-cellular, rhizoidal filaments that penetrate the polysaccharide material of the host. Erect filaments unbranched, 1–2 mm tall, about 10 μm in diameter at the base, increasing upward to 35–50 μm in diameter. Uniseriate below but pluriseriate above in mature plants.

Reported from Virginia by Rhodes (1970a), but the host plant not mentioned. Known from southern Massachusetts as an epiphyte on *Porphyra.* Probably much more widely distributed than present records indicate.

Cleland in Collins, 1918, p. 144, pl. 124, figs. 6–9; Taylor, 1957, p. 203.

GENUS *ERYTHROCLADIA*

Plants completely prostrate, of spreading, branched filaments or forming compact disks of pseudodichotomously branched filaments with the
marginal cells bifid; epiphytic, endophytic, on animals, or on nonliving
substrata. Chromatophore a lobed, parietal plate with one pyrenoid.
Reproduction by monospores produced by the older cells.

ERYTHROCLADIA SUBINTEGRA ROSENVINGE

Plants producing a compact disk of one cell layer, the marginal cells
bifid before division, the branching pseudodichotomous and the filaments
radiating from the center. Cells in older plants becoming pseudoparenchymatous in arrangement. Cells mostly 4–6 μm in diameter.

On various algae, York River at Gloucester Point and the lower
Chesapeake Bay in general. Year-round. Known from Brazil through
the tropical Atlantic to Massachusetts.

Rosenvinge, 1909, p. 73, figs. 13–14; Børgesen, 1916–20, p. 7, figs.
3–4; Collins and Hervey, 1917, p. 95; Taylor, 1960, p. 290, pl. 41,
fig. 1.

GENUS *ASTEROCYTIS*

Plants filamentous, uniseriate, branched, grayish-green to purplish-green.
Cells ellipsoid or oval with a central, stellate chromatophore. Asexual
reproduction by monospores.

ASTEROCYTIS RAMOSA (THWAITES) GOBI

Plants to a few millimeters tall, alternately or subdichotomously
branched, usually grayish-green. The filaments are 12–20 μm in diameter, with thick gelatinous walls. The cells are ovate to oblong, 5–8 μm
in diameter, 8–20 μm long. Vegetative cells become akinetes or monospores and escape from the filament through the side or end of the
gelatinous wall.

Known from the Caribbean Sea to Nova Scotia and Newfoundland
as an epiphyte on seagrass leaves and other algae. Apparently not yet
reported in Virginia, although it surely occurs there.

Taylor, 1928, p. 132, pl. 20, figs. 1–2; Taylor, 1957, p. 201; Humm
and Hildebrand, 1962, p. 249 (as *A. ornata* [C. Agardh] Hamel);
Humm, 1964, p. 309; Mathieson, Dawes, and Humm, 1969, p. 130.

GENUS *GONIOTRICHUM*

Plants of erect, usually pseudodichotomously branched filaments with thick longitudinal walls, uniseriate or becoming pluriseriate. Cells short, with an axial radiating chromatophore and one pyrenoid, the nucleus to one side. Reproduction by formation of monospores released by dissolution of a side of the cell wall.

GONIOTRICHUM ALSIDII (ZANARDINI) HOWE
(Fig. 29)

Plants 0.5–5.0 μm tall, rose-red to greenish, usually much-branched, the filaments 12–20 μm in diameter, the cells 4–13 μm long, depressed-spherical to ellipsoid or cylindrical.

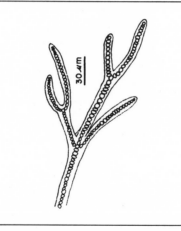

Fig. 29.   *Goniotrichum alsidii,* the upper part of a plant

Common as an epiphyte in the lower Chesapeake Bay and along the Eastern Shore of Virginia during summer and early fall. Known from the Caribbean Sea to Newfoundland, usually as an epiphyte on seagrasses and larger algae in quiet waters.

Howe, 1914, p. 75; Børgesen, 1916–20, p. 4, fig. 2 (as *G. elegans* [Chauvin] Lejolis); Hoyt, 1917–18, p. 465, fig. 23; Humm and Caylor, 1957, p. 250, pl. 7, fig. 4; Taylor, 1957, p. 202, pl. 28, figs. 1–4; Humm and Hildebrand, 1962, p. 249; Kim, 1964, p. 128; Mathieson, Dawes, and Humm, 1969, p. 131.

ORDER BANGIALES

Plants of two morphological forms, a filament that becomes pluriseriate, and a flat sheet one or two cells thick. These in turn have two morphologically different growth forms that are self-reproducing or that produce the other form in response to environmental conditions. Several types of spores are formed, but sexual reproduction, if it occurs in this order, is not understood. Cells with stellate chromatophores having a central pyrenoid.

FAMILY BANGIACEAE

With characters of the order.

GENUS *BANGIA*

Plants producing two stages. In the one traditionally known as *Bangia,* the plants are erect, filiform, unbranched, at first uniseriate but becoming pluriseriate; attached by a holdfast cell that is supplemented by rhizoidal downgrowths from cells just above it. Chromatophores axial, with radiating lobes. Reproduction occurs by production of monospores and vegetatively by formation of bipolar germlings.

In the stage of the plant known as the conchocelis stage (formerly the genus *Conchocelis*), the plants are uniseriate, branched filaments that often penetrate oyster shells and plates of barnacles. *Conchocelis* arises from the bangia stage, at least in culture, whenever *Bangia* is subjected to twelve hours or more of light per day. *Bangia* arises from the conchocelis stage when there is less than twelve hours of light per day through the formation of "fertile cell rows" that produce plantlets. Under longer periods of illumination, *Conchocelis* produces monospores that germinate into *Conchocelis* plants. There is evidence, however, that this behavior in culture is not duplicated in nature, but that considerable variation exists (Dixon and Richardson, 1969), as the natural environment is different from any laboratory culture environment. Basal portions of the bangia stage may persist through the adverse season of the year, where this stage is seasonal, and give rise to *Bangia* plants the following year.

BANGIA FUSCOPURPUREA (DILLWYN) LYNGBYE

Plants (in the bangia stage) of attached filaments, 2–10 cm long (or more), 20–200 μm in diameter, forming brownish-red patches or a band high in the intertidal zone. Filaments at first uniseriate, the cells shorter than, or equal to, the diameter, later dividing radially and producing thick, pseudoparenchymatous filaments, especially in the upper parts.

Conchocelis-stage plants of uniseriate, branched filaments about 5–10 μm in diameter and usually penetrating the shells of oysters, barnacles, or other molluscs; sometimes developing on the surface of stones or solid objects. The conchocelis stage is present throughout the year and most common in the intertidal zone, but it also occurs below low tide to a depth of at least one meter. It is often difficult to find in nature.

The distribution notes and references that follow refer only to the bangia stage.

Texas, Bermuda, South Carolina to Newfoundland. In the southern part of its range, it is a plant of winter and spring. It is best developed in the northern part of its range. In Virginia it has been observed to form an intertidal band on pier pilings near the Virginia Institute of Marine Science at Gloucester Point and also near the laboratory at Wachapreague and to persist as late as July. Of general distribution in the lower Chesapeake Bay and along the Eastern Shore.

Lyngbye, 1819, p. 83, pl. 24; Hoyt 1917–18, p. 464; Taylor, 1957, p. 204, pl. 28, figs. 10–12; Earle and Humm, 1964; Wulff and Webb, 1969; Edwards, 1970, p. 30, figs. 98–101.

GENUS *PORPHYRA*

Plants a thin sheet one or two cell layers thick, attached by a group of rhizoids arising as extensions of the basal cells. "Alpha" spores of *Porphyra* produce branched, uniseriate, filamentous plants (the conchocelis stage) that usually penetrate shells and are perennial. This stage produces conchospores that germinate to produce the leafy *Porphyra* plants (Chen et al., 1970). In the past, the conchocelis stage was placed in the genus *Conchocelis*, family Acrochaetiaceae, order Nemaliales. "Beta" spores of the leafy (*Porphyra*) stage reproduce this stage. The leafy stage is sometimes perennial, but more often it is seasonal. In the southern part of its range (Virginia to Cape Canaveral, Florida), the plants usually appear late in November and disappear about mid-April. Small, basal portions of the leafy plants persist, however, and give rise to new upper parts the following season (Dixon and Richardson, 1969).

Coll and Cox (1977) recently completed a study of the genus *Por-*

phyra at Beaufort, North Carolina, and came to the conclusion that the *Porphyra* plants in that area represented two species that had not been described. Accordingly, they described *P. carolinensis* and *P. rosengurt-tii*.

Since the work of Hoyt (1917–18) and Williams (1948) at Beaufort, two species were known for the area, *P. leucosticta* and *P. umbilicalis*. Plants assigned to the former usually appear each fall in November or December and disappear in April. Plants assigned to the latter behaved in a similar manner except that, since about 1960, the porphyra stage of the latter began to persist the year around on large rocks and on the breakwater at Fort Macon in a dwarf form 1–2 cm broad. These plants could, however, represent an introduction of another species into the area about twenty years ago.

If Coll and Cox are correct in their interpretation of the taxonomy of *Porphyra* at Beaufort, then the same errors have probably been made along the Virginia coast, and their new species are present in Virginia as well. Both are monostromatic. *P. carolinensis* has small, marginal teeth; the margins of *P. rosengurttii* are entire.

Coll and Cox expressed the opinion that the reason so few species of *Porphyra* are known for the Atlantic coast of North America is because of a lack of detailed examination of *Porphyra* from that region.

Key to the Species of the Genus *Porphyra*

1.   Plants rose-red, of two layers of cells . . . . .   *P. miniata*

1.   Plants purplish-red to brownish-red, one cell thick 2

2.   Spores in longitudinal patches in outer parts of the blade . . .
. . . . *P. leucosticta* (p. 68)

2.   Spores produced in a marginal band . . . .
. . . . *P. umbilicalis* (p. 68)

PORPHYRA MINIATA (LYNGBYE) C. AGARDH

Plants rounded, ovate, or oblong, sometimes wider at the base, 15–30 cm in diameter when well developed. The blade is at first one cell thick at the margins but becomes two cells thick (30–70 μm), the cells isodiametric in section. Spores are produced along the margins. The life history of this species has been worked out by Chen et al. (1970).

Known from Virginia to the Arctic. While primarily a winter-spring species in Virginia, a collection made in July 1962 suggests that it may persist through the summer in the leafy form. It is intertidal on oyster shells, stones, seawalls, and woodwork.

Lyngbye, 1819, p. 29, pl. 6, fig. d (as *Ulva miniata*); C. Agardh, 1824, p. 191; J. Agardh, 1873–90 (1882), p. 59, pl. 2, figs. 44–48; Chen et al., 1970, figs. 1–27.

PORPHYRA LEUCOSTICTA THURET

Plants more or less oblong in shape, about 8–15 cm in height, 6–12 cm in width, attached by a group of rhizoids. Blades 25–50 μm thick, of a single layer of cells that are 1.5 to 2.0 times as high as they are wide, about 12–15 μm in diameter in surface view. Spores produced in pale, elongated patches that are parallel to each other near the margin of the blade, 1.0–1.5 mm wide, 5.0–10 mm long.

On intertidal oyster shells, rocks, and woodwork in Virginia, apparently present in the leafy form only from November to May. This species is known from intertidal rocks along the ocean beach north of Cape Canaveral, Florida, to the Arctic. In the Gulf of Mexico, it is known only from the Port Aransas, Texas, area.

Thuret in LeJolis, 1863, p. 100; J. Agardh, 1883, p. 64, pl. 2, figs. 55–58; Hoyt, 1917–18, p. 465; Humm, 1952; Taylor, 1957, p. 206; Taylor, 1960, p. 295; Edwards, 1970, p. 31, figs. 102–5.

PORPHYRA UMBILICALIS (L.) J. AGARDH
(Fig. 30)

Plants oblong to strap-shaped, narrow when young but becoming broad and rounded at the base, the upper part often lobed, generally 8–20 cm tall, 4–8 cm or more in diameter, greenish- or brownish-purple. Blades one cell (30–75 μm) thick. Cells 8–25 μm in diameter in surface view, isodiametric or somewhat higher than they are wide as viewed in vertical section. Spores are produced in light-colored, often ragged, marginal bands, or sometimes scattered over the surface.

In Virginia, this species is common on oyster shells and barnacles in the intertidal zone, and on other substrata, from November to April or May. It has been reported as occurring in the summer in the leafy form (Zaneveld and Barnes, 1965) and may persist throughout the summer in some localities.

The species is known from South Carolina (Wiseman, 1966) to the Arctic. In the northern part of its range, the leafy form is primarily a summer plant; south of Cape Cod it is best developed during winter and spring. The conchocelis stage is perennial.

Fig. 30. *Porphyra umbilicalis,* photograph of an herbarium specimen

J. Agardh, 1883, p. 66, pl. 2, fig. 61; Williams, 1948, p. 690; Taylor, 1957, p. 206; Wulff and Webb, 1969.

SUBCLASS FLORIDEOPHYCIDAE

Most members of the Florideophycidae, with the exception of the order Nemaliales, produce gametophytes and sporophytes that are isomorphic. Growth is terminal. While many genera are pseudoparenchymatous in appearance, all are regarded as fundamentally filamentous. Daughter cells retain protoplasmic connections that are usually readily visible, and these aid in tracing filaments and cell derivatives.

Reproductive structures and processes are relatively complex. The carpogonium, or egg cell, is borne terminally on a lateral filament, the carpogonial branch, and produces an extension, the trichogyne. Fertilization follows contact of a spermatium with the trichogyne. The zygote then produces, without meiosis, a few-celled carposporophyte, and this in turn produces many carpospores, each of which is genetically a duplicate of the zygote. In some orders of the Florideophycidae, the zygote nucleus is first transported from the carpogonium to a proximate cell of the female plant, the auxiliary cell, through a special filament, the ooblast. The auxiliary cell is then the center of development of the carposporophyte. The carposporophyte gives rise to a cluster of filaments, the gonimoblasts, and these produce a series of carpospores at their tips. Cells of the female plant in the immediate vicinity of the auxiliary cell may serve as nurse cells or nurse tissue in which food reserves accumulate. The proximate female tissue may also give rise to a vase-shaped layer of cells that surround the carposporophyte, the pericarp. The pericarp and carposporophyte constitute the cystocarp.

Orders of the Florideophycidae have been established on the basis of the mode of development of the cystocarp. The process is obscure, but no better means has been found for subdividing the Florideophycidae into orders of apparent phylogenetic significance.

Carpospores germinate into sporophyte plants that in turn produce spores in groups of four, following meiosis. These are the tetraspores. Culture experiments have shown that two of a group of tetraspores will produce male plants, the other two, female plants. The majority of the Florideophycidae are dioecious. Nearly all are marine.

ORDER NEMALIALES

Plants minute to relatively large, of free filaments or corticated; cells uninucleate, with protoplasmic connections; chromatophores central or parietal, single or numerous. Asexual reproduction by monosporangia, rarely by bisporangia or tetrasporangia. Sexual reproduction by spermatia and carpogonia, the zygote said to undergo meiosis, producing the carpospores directly.

In recent decades, an increasing number of species placed in this order have been shown to have both gametophyte and sporophyte stages. It may be that the order, as presently defined, cannot be maintained.

FAMILY ACROCHAETIACEAE

Plants minute, to about 6 mm maximum height, of branched filaments with apical growth, prostrate or erect, or with both prostrate and erect filaments, uniseriate. Reproduction usually by monosporangia, but bisporangia or tetrasporangia are sometimes produced. Gametophyte plants may produce spermatia on short, branched, photosynthetic filaments or on colorless filaments or carpogonia that are terminal on one-celled or two-celled branches. Cystocarps produce carpospores from sparingly branched gonimoblast filaments.

The family is treated here *sensu* Rosenvinge (1909) and Aziz (1965). The latter, following a study of western Atlantic species of this family, recognized only two genera, *Kylinia* and *Acrochaetium*.

GENUS *ACROCHAETIUM*

Plants minute (to about 5 mm tall, but usually less than 2 mm), epiphytic or endophytic, rarely on shells or stones; filamentous, uniseriate, branched. The basal portion may consist of a single cell (modified original spore) or of one or a group of creeping filaments. Cells uninucleate, with one axial stellate or one parietal chromatophore with or without a pyrenoid. Some plants have parietal chromatophores in the upper parts, stellate chromatophores in the basal parts or in the monospores. Cells of the creeping filaments may have several discoid chromatophores.

Asexual reproduction by monosporangia (usually) that are lateral or terminal, sessile or stalked. Bispores and tetraspores are known for some species.

Sexual reproduction apparently uncommon, involving spermatia produced in clusters or in pairs and solitary carpogonia that are lateral or terminal, sessile or stalked. Gonimoblast filaments of a few cells give rise to several carpospores. Plants monecious or dioecious.

The key that follows is based upon the studies of Aziz (1965).

Key to the Species of the Genus *Acrochaetium*

1. Endozooic in the colonial, cartilaginous bryozoan *Alcyonidium* . . . . *A. alcyonidii*

1. Not endozooic 2

2. Base one-celled, consisting of the original spore . . . . *A. trifilum* (p. 74)

2. Base of multicellular filaments or of a disk of filaments . . . 3

3. Base of free filaments, though sometimes aggregated and disk-like 4

3. Base a parenchymatous disk, sometimes with free marginal filaments 8

4. Base entirely endophytic; original spore distinct, producing one erect filament . . . . *A. dasyae* (p. 74)

4. Base entirely external to host or only partly penetrating . . . 5

5. Base entirely external to host 6

5. Base partly penetrating the host 7

6. Monosporangia 14–16 μm long, not secund . . . . *A. flexuosum* (p. 75)

6. Monosporangia 18–22 μm long, often in secund series . . . . *A. sagraeanum* (p. 75)

7. Base a pseudoparenchymatous disk with a thick-walled endophytic filament about 20 μm long . . . . . *A. robustum* (p. 76)

7. Base of loosely branched filaments; monosporangia 8–12 μm in diameter . . . . *A. daviesii* (p. 76)

8. Chromatophores axial, stellate . . . *A. virgatulum* (p. 77)

8. Chromatophores parietal, not stellate . . *A. thuretii* (p. 77)

ACROCHAETIUM ALCYONIDII JAO
(**Fig. 31**)

Plants growing within the gelatinous matrix of *Alcyonidium verrilli* and producing short, external branches bearing monosporangia and lending

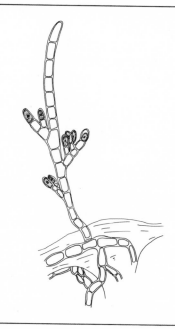

Fig. 31. *Acrochaetium alcyonidii.* Filaments embedded in the gelatinous matrix of the bryozoan host, *Alcyonidium verrilli,* have given rise to a free, erect branch bearing monospores.

a distinct reddish color to the host. Erect, exserted filaments mostly 100–600 μm tall, of somewhat greater diameter than the endozoic filaments, which range from 5–13 μm in diameter but are mostly 5–6 μm. Asexual reproduction by monospores (rarely by bispores). Sexual reproduction unknown.

The filaments ramify through the gelatinous surface layer of the host and also penetrate to the center between the zooids. This deep penetration of the host and the fact that the plant does well on hosts growing attached in moderately deep water where light penetration is poor strongly suggest that the plant obtains organic matter from the host. It may produce an enzyme capable of hydrolysis of a component of the host gelatinous material. It presents a problem that deserves investigation.

A. alcyonidii was collected at Cape Charles, Virginia, in 1946. During the summer of 1962 it was abundant in the Chesapeake Bay, especially the Hampton Roads area. It was originally described from Massachusetts and has been recorded from Nova Scotia on the hydroid, *Sertularia.*

Jao, 1936, p. 245; Taylor, 1957, p. 220, pl. 34, figs. 2–4; Aziz, 1965, p. 89, pl. 13, fig. 4; Edelstein, McLachlan, and Craigie, 1967, p. 195.

ACROCHAETIUM TRIFILUM (BUFFHAM) BATTERS, EMENDED AZIZ

Plants attached by a single basal cell (the original spore) and reaching a height of 15–200 μm. The basal cell may be on the surface of the host or embedded in the gelatinous outer layer; it is spherical to oblong, 5–15 μm in diameter or length, and it gives rise to one to six erect filaments that are branched or unbranched. Erect filaments 3–15 μm in diameter, the cells one to two times as long as broad, the filaments tapering upwards and with or without terminal, unicellular hairs. Chromatophores usually stellate and with one pyrenoid. Monosporangia 6–9 μm in diameter, 7–16 μm long, spherical to oval, and usually sessile, though sometimes pedicellate. Sexual reproduction not common. Spermatia produced at the tips of short, dichotomous branchlets; carpogonia about 6 μm in diameter at the base, about 19 μm long, the trichogyne about 11 μm long, 2–3 μm in diameter; carpospores oval, 8–10 μm in diameter, 12–15 μm long, terminal.

The above description is abbreviated from that of Aziz (1965), whose study of the genus led him to the conclusion that *Acrochaetium crassipes* (Børgesen) Børgesen (1916–20), *A. compactum* Jao (1936), and *A. parvulum* Hoyt (1917–18) are synonyms. Aziz studied hundreds of specimens and decided that *A. trifilum* was the oldest valid name; accordingly, he emended the description to include twenty-one species names that he considered to be synonyms (some of which are in the genus *Kylinia, sensu* Papenfuss, 1945, 1947).

The species is widely distributed and moderately variable. It was first recognized in Virginia from a collection made by Dr. K. M. S. Aziz at Haven Beach, August 4, 1962, where it occurred on *Zostera*. It is known from the Caribbean Sea to Newfoundland.

Batters, 1902, p. 58; Børgesen, 1916–20, p. 20 (as *A. crassipes*); Jao, 1936, p. 241; Williams, 1948, p. 690 (as *A. compactum*); Aziz, 1965, p. 34, pl. 2, figs. 1–16, and pl. 11, figs. 1–7; Mathieson, Dawes, and Humm, 1969, p. 132 (as *Kylinia compacta*).

ACROCHAETIUM DASYAE COLLINS

Plants with the basal portion embedded in the tissue of *Dasya baillouviana*, the only host on which it has been reported, apparently, throughout its extensive range. Plants reaching a height of about 2 mm, the basal portion consisting of the original spore, from which arises a downward penetrating filament and a single erect filament. Erect filaments 6–8 μm in diameter, branched above, and bearing monosporangia about 10–15 μm in diameter and 20–25 μm long. Aziz (1965) was the first to report sexual plants, which he found in North Carolina in April 1963 and 1964, and which he describes in detail.

The species was first collected in Virginia by Katrine deWitt on *Dasya baillouviana* in Broad Bay, Virginia Beach, April 10, 1949, and was first recognized by Aziz in 1965. Since he regards *A. intermedium* Jao (1936) as a synonym, the species is known from Nova Scotia, Massachusetts, Maryland (Point Lookout), North Carolina (Beaufort), and the Gulf coast of central Florida (mouth of the Homosassa River). During the winter months it is probably continuous from North Carolina to the West Indies, but in its northern range it may be present only during the warmer months.

Collins, 1906, p. 191; Jao, 1936, p. 242 (as *A. intermedium*); Taylor, 1957, p. 217; Aziz, 1965, p. 76, pl. 6, figs. 1–2, and pl. 7, figs. 1–5; Edelstein, McLachlan, and Craigie, 1967, p. 195, fig. 9.

ACROCHAETIUM FLEXUOSUM VICKERS

Plants in dense tufts usually about 0.7 mm high (to 2.0 mm) arising from an entangled base of creeping filaments, which may be closely associated toward the center to form a pseudoparenchymatous disk. Erect filaments much-branched, 6–10 μm in diameter; the cells 2–5 diameters long. Chromatophores parietal, with a conspicuous pyrenoid. Monosporangia ovate to oblong, on the upper sides of branchlets, 9–10 μm in diameter, 14–16 μm long.

In Virginia this species has been found in abundance on *Ceramium rubrum* and *Gracilaria foliifera* from the breakwater at Cape Charles and from several stations along the Eastern Shore, especially at Wachapreague. Known from the Caribbean Sea to Newfoundland.

Vickers, 1905, p. 60; Børgesen, 1916–20, p. 34, figs. 29–30; Taylor, 1957, p. 220; Aziz, 1965, p. 66, pl. 13, figs. 1–2.

ACROCHAETIUM SAGRAEANUM (MONTAGNE) BORNET

Plants in tufts or in dense stands 2–5 mm tall, arising from a base of entangled filaments. Erect filaments about 12 μm in diameter below, with cells 4–8 diameters long and tapering upwards. Monosporangia 8–10 μm in diameter, 18–22 μm long, one or two arising from each cell of short branchlets. Tetrasporangia occasionally produced, 17–25 μm in diameter, 28–34 μm long.

On *Gracilaria foliifera* from the Chesapeake Bay at Little Creek near Norfolk, Virginia, June 18, 1974, collected and determined by Dr. F. D. Ott. Known from the Caribbean Sea to Connecticut.

Montagne, 1856 (as *Cladophora sagraeana*); Bornet, 1904; Humm and Hildebrand, 1962, p. 250; Taylor, 1957, p. 220; Taylor, 1960, p. 315; Aziz, 1965, p. 66, pl. 13, figs. 1–2; Humm and Hamm, 1976, p 43; Hamm and Humm, 1976, p. 213; Humm, 1976, p. 76.

ACROCHAETIUM ROBUSTUM BØRGESEN

Plants growing to a height of 1.0–1.5 mm from a basal layer of closely appressed filaments that form a pseudoparenchymatous disk. From the original spore, a single, thick-walled filament penetrates the host below the disk. Erect filaments numerous, sparingly branched, 9–10 μm in diameter. Cells thick-walled, 1.5–2.0 diameters long below and to 4 or 5 diameters long above. Monosporangia about 11 μm in diameter, 14–16 μm long, the walls to 2 μm thick. Sexual reproduction unknown.

Known from Virginia and North Carolina and from the Caribbean Sea. It seems strange that this species has not been reported for Florida. The original collection in Virginia was made by Michael Castagna along the inner shore of Cedar Island near Wachapreague, November 2, 1962, and the specimens were determined by K. M. S. Aziz (1965). *Gracilaria foliifera* was the host plant.

Børgesen, 1916–20, p. 40, figs. 38–40; Pearse and Williams, 1951, p. 153; Taylor, 1960, p. 315; Aziz, 1965, p. 68, pl. 13, fig. 3.

ACROCHAETIUM DAVIESII (DILLWYN) NÄGELI

Plants to a height of 6 mm from a base of creeping filaments that may become rather dense and entangled. Erect filaments much-branched, especially along one side, 7–13 μm in diameter below, somewhat less above. Cells 2–4 diameters long and with a single parietal chromatophore having one pyrenoid. Branches usually terminating in slender, multicellular, hairlike tips. Monosporangia borne laterally, sessile or short-stalked, 8–12 μm in diameter, 10–20 μm long, ovate to oblong. Tetrasporangia sometimes produced, 7–12 μm in diameter, 13–16 μm long.

Aziz (1965) reported sexual reproduction on specimens collected in Jamaica in 1962. Carpospores are borne terminally on the gonimoblast filaments and are about 10 × 14 μm.

Collected in Virginia from Burtons Bay near Wachapreague on old stems of *Spartina* in a tidal pool by Dr. F. D. Ott, August 15, 1972. Known from the Caribbean Sea and New England. It is to be expected along this entire coastline.

Dillwyn, 1802–9, p. 73 (as *Conferva daviesii*); Nägeli, 1861, p. 405; Collins, 1906, p. 194; Taylor, 1957, p. 221, pl. 31, figs. 8–10; Aziz, 1965, p. 70.

ACROCHAETIUM VIRGATULUM (HARVEY) J. AGARDH

Plants arising from a parenchymatous basal disk that produces many erect filaments to a height of 1–2 mm. Erect filaments branched, 7–14 μm in diameter, and often with unicellular, terminal hairs. Cells 2–5 diameters long with chromatophores that are usually axial and stellate and with one pyrenoid. Monosporangia lateral and sessile, or terminal on short branchlets, ovoid, 10–18 μm in diameter, 12–22 μm long. Tetrasporangia occur rarely. Sexual reproduction has not been reported.

While not yet reported from Virginia, this species is known from North Carolina, New York, New England, Nova Scotia (collected by Katrine deWitt), and Newfoundland. It is a summer species in New England and northward.

Harvey, 1833, p. 349 (as *Callithamnion virgatulum*); J. Agardh, 1892, p. 48; Hoyt, 1917–18, p. 473, figs. 29–30; Aziz, 1965, p. 90, pl. 12, fig. 3, and pl. 15, fig. 4; Mathieson, Dawes, and Humm, 1969, p. 132 (as *Kylinia virgatula*).

ACROCHAETIUM THURETII (BORNET) COLLINS AND HERVEY

Plants arising from an irregular, parenchymatous basal disk that is 60–120 μm in diameter. Erect branches numerous, 2–3 mm high, the main filaments 8–12 μm in diameter, the branching mainly from the lower parts. Cells 3–5 diameters long below, 8–12 diameters long above, containing a single, parietal, platelike chromatophore with one pyrenoid. Monosporangia usually on one-celled stalks, 9–11 μm in diameter, 14–17 μm long, ovate. Sexual plants monoecious, the spermatia and carposporangia borne near the base of branches like the monosporangia; the carposporangia 11–14 μm in diameter, 18–21 μm long.

Known in Virginia by a single collection from brackish water in the lower James River by William S. Johnson, but it is to be expected at many localities along the Virginia coast. Known from Florida, Bermuda, North Carolina, and New England.

Bornet, 1904 (as *Chantransia efflorescens,* var. *thuretii*) ; Collins and Hervey, 1917, p. 97; Taylor, 1957, p. 222; Taylor, 1960, p. 310; Aziz, 1965, p. 93; Ballantine and Humm, 1975, p. 156; Humm, 1976, p. 76.

ORDER GELIDIALES

Plants tough and wiry, often purplish-red to blackish, tenaciously attached to the substratum, sparingly to densely branched, growing from an apical cell or a group of apical cells and producing the multiaxial type of structure with a corticating layer. Reproduction by tetraspores immersed or at the surface and usually in bands or groups. Sexual reproduction of gametophytes characterized by direct development of the cystocarp from the fertilized carpogonium, with carpospores produced terminally on gonimoblast filaments.

There is only one family, Gelidiaceae, and about six genera.

GENUS *GELIDIUM*

Plants growing from an apical cell that produces an axial row and these giving rise to pericentral cell rows; later the pericentrals are paralleled by many additional medullary filaments of secondary origin, the filamentous core covered by short, compact, radially branched cell series making up a photosynthetic cortex. Thick-walled filaments (rhizines) are present in the axis among the larger cell rows, either in the inner cortex or somewhat scattered, but not restricted to the medulla. Sporophyte plants produce tetraspores in tetrapartite arrangement in definite zones of the branchlets. Sexual reproduction is characterized by the absence of an auxiliary cell, the carposporophyte developing directly from the carpogonium. The cystocarp is usually bilocular because of a median septum and usually has two openings. The plants are dioecious.

GELIDIUM CRINALE (TURNER) LAMOUROUX

(Fig. 32)

Plants producing primary creeping branches that spread over the substrate and give rise to tufts of erect branches 2–8 cm tall, alternately branched below, usually pinnately branched above, the branchlets often flattened, especially in tetrasporic plants. Mature tetrasporangia 10–22

Fig. 32. *Gelidium crinale,* photograph of an herbarium specimen (×2)

μm in diameter, 13–25 μm long, in zones in the branchlets. Cystocarps in enlarged areas of the branchlets, solitary or in pairs.

Common in the Chesapeake Bay and along the Eastern Shore of Virginia on shells and stones low in the intertidal zone or just below. Known from the Caribbean Sea to the Gulf of St. Lawrence.

Turner, 1808–19 (1819), pl. 198 (as *Fucus crinalis*); Lamouroux, 1825, p. 191; Farlow, 1881, p. 158; Hoyt, 1917–18, p. 475, pl. 95, fig. 2; Taylor, 1957, p. 231, pl. 35, figs. 1–3, pl. 40, fig. 3, and pl. 41, fig. 5; Edwards, 1970, p. 33, figs. 115–19.

ORDER CRYPTONEMIALES

Plants of uniaxial or multiaxial structure and exhibiting a wide variety of form, some heavily calcified. Reproductive structures free or produced in conceptacles. In this order the ooblast fuses with an auxiliary cell borne in an intercalary position on a special filament either upon the supporting cell of the carpogonial filament or from a vegetative cell.

FAMILY CORALLINACEAE

Plants heavily calcified, of multiaxial growth, and of a wide variety of forms. Tetrasporangia, antheridia, and carposporophytes borne in separate conceptacles with one or more pores. Following fertilization of the carpogonium, zygote nuclei are transferred to the auxiliary cell. Fusions with other cells occur, producing a large cell from which gonimoblast filaments arise, these producing carpospores.

SUBFAMILY MELOBESIEAE

Plants forming a thin or massive crust with or without erect branches, calcified throughout.

GENUS *MELOBESIA*

Plants consisting of a thin calcified crust of one cell layer in the vegetative parts, but each cell developing a small superficial cell; three to five cell layers thick in the vicinity of the conceptacles. Tetrasporangial conceptacles with several pores. Plants monoecious or dioecious.

MELOBESIA MEMBRANACEA (ESPER) GREVILLE

Plants pale reddish-purple or white, forming thin crusts on the leaves of seagrasses and larger algae. Cells, as seen from above, 5–9 μm in diameter, 9–20 μm long. Tetrasporangial conceptacles with eight to twenty-seven pores, from 100–150 μm in diameter, scattered, or grouped and confluent. Spermatangial and carpogonial receptacles hemispherical, with a single pore, the aperture contracted by filiform cells from the internal walls.

In Virginia on *Zostera* leaves, on *Gracilaria foliifera* and other algae, from the breakwater at Cape Charles and along the Eastern Shore.

Known from the Caribbean Sea to northern Massachusetts, a summer plant in the northern part of its range but present the year around in the south.

Lamouroux, 1812, p. 186; Foslie, 1929, p. 49 (as *Lithothamnium membranaceum*); Taylor, 1957, p. 250; Humm, 1964, p. 310; Humm, 1976, p. 77.

GENUS *FOSLIELLA*

Plants similar to *Melobesia,* but the tetrasporangial conceptacles have but one pore and are more conical in shape. Cystocarpic conceptacles are also similar but smaller.

Key to the Species of the Genus *Fosliella*

Enlarged, colorless cells terminating cell rows in the surface of the plant . . . . *F. farinosa*

Distinctive cells as above not present . . . . *F. lejolisii*

FOSLIELLA FARINOSA (LAMOUROUX) HOWE

Plants of thin fragile crusts, mostly 2–5 mm in diameter. Vegetative cells 8–15 μm wide, 12–30 μm long (as seen in surface view), each row terminated by a colorless, swollen, hair-bearing cell (trichocyte) 15–30 μm in diameter, 20–40 μm long. Sporangial and cystocarpic conceptacles hemispherical to conical, protruding considerably above the surface of the crust, and having a single apical pore, 150–250 μm in diameter. Spermatangial conceptacles smaller, 60–80 μm in diameter. Tetrasporangia 30–50 μm in diameter, 50–90 μm long.

Widely distributed in Virginia marine waters as an epiphyte on larger algae and on *Ruppia* and *Zostera,* especially along the Eastern Shore, where growth is relatively rapid and reproduction occurs during the warmer months. Known from the Caribbean Sea to New England.

Lamouroux, 1816, p. 315, pl. 12, fig. 3 (as *Melobesia farinosa*); Howe, 1920, p. 587; Taylor, 1957, p. 252; Humm, 1964, p. 311; Ballantine and Humm, 1975, p. 157.

FOSLIELLA LEJOLISII (ROSANOFF) HOWE

Plants of thin fragile crusts, mostly 0.5–2.0 mm in diameter, only 15–30 μm thick, and consisting of coalesced, radiating filaments 6–7 μm wide, the cells 6–10 μm long, as seen in surface view. Cells that bear a fork are somewhat larger. Plants one cell thick near the margin or with scattered, small, superficial cells; two to four cells thick at the center. Where the crust is three cells thick, the central layer of cells is much taller (15–20 μm) than the upper or lower layer (2–4 μm). Sporangial and cystocarpic conceptacles convex, often numerous and crowded, 150–250 μm in diameter; tetrasporangia 30–50 μm in diameter, 50–80 μm long. Spermatangial conceptacles 75–100 μm in diameter.

Foslie (1909) established the genus *Heteroderma* for those members of the subfamily Melobesieae with trichocytes. Mason (1953) and Dawson (1960) both distinguished between *Fosliella* and *Heteroderma*. Since most Atlantic coast phycologists have not recognized *Heteroderma*, this species has been retained in *Fosliella* here, in agreement with the familiar literature dealing with the algae of the western Atlantic.

As pointed out by Taylor (1960), Lemoine (1917) believes that specimens of *F. lejolisii* from the southern states and the tropics are a form of *F. farinosa* with few or no trichocytes. This problem awaits a careful study of the group. But if the presence or absence of trichocytes is as variable as Mme. Lemoine suggests, then the genus *Heteroderma* may not be defensible.

In Virginia, *F. lejolisii* occurs on leaves of *Zostera* along the Eastern Shore, at Cape Charles, at Wachapreague, and elsewhere. It has been recorded from the Caribbean Sea to the Canadian maritime provinces.

Rosanoff, 1886, p. 62, pl. 1, figs. 1–13; (as *Melobesia lejolisii*); Foslie, 1909, p. 56 (as *Heteroderma lejolisii*); Howe, 1920, p. 588; Dawson, 1960, p. 55, pl. 50, figs. 4–6 (as *Heteroderma lejolisii*); Humm, 1964, p. 311; Humm, 1976, p. 77; Chamberlain, 1977, p. 73, figs. 14–22.

SUBFAMILY CORALLINEAE

Plants erect, branched, arising from a basal crust. Erect branches divided into long and short segments, the long segments calcified, the short segments not calcified. These are the articulated corallines.

GENUS *CORALLINA*

Plants consisting of calcareous, articulating segments in the erect branches, the branching usually pinnate, at least in the upper parts, the

branchlets often opposite. Erect branches terete to flattened. Conceptacles on the branchlets, with or without hornlike projections and with
an apical pore.

CORALLINA OFFICINALIS LINNAEUS

Plants in tufts, usually 5–8 cm tall, much-branched, the branches pinnate
and in one plane. Segments of the branches terete in the lower parts
and flattened in the upper parts. Conceptacles ovate to subspherical,
only the spermatangial conceptacles having hornlike projections.

C. officinalis is a variable species, plants from shallow water differing
in several minor respects from those that grow in deep water. There
are differences also between those from New England and from the
southern part of the range, North Carolina and South Carolina.

Though the species apparently has not been found in Virginia waters,
it may occur in relatively deep water off the Eastern Shore or on solid
surfaces off Cape Henry. Since it is stenohaline, it apparently does not
grow in the Chesapeake Bay. It is known from South Carolina to the
maritime provinces of Canada and from Bermuda.

Linnaeus, 1761, p. 539; Harvey, 1853, p. 83; Farlow, 1881, p. 179;
Williams, 1948, p. 691; Taylor, 1957, p. 254, pl. 36, figs. 1–5; Taylor,
1960, p. 410.

GENUS JANIA

Plants with erect branches arising from a small, basal disk. Branching
dichotomous, the branches segmented, calcified. Conceptacles solitary in
swollen terminal segments and having an apical pore, usually with a pair
of branches arising from just below the conceptacle.

JANIA CAPILLACEA HARVEY

Plants 5–8 mm tall, of slender, dichotomous branches with wide angles,
50–100 μm in diameter, the segments four to six times as long as wide.

This species is not part of the Virginia flora but is often found as an
epiphyte on the two pelagic species of Sargassum that wash ashore along
the outer beaches.

Harvey, 1853, p. 84; Hoyt, 1917–18, p. 527 (as Corallina capillacea); Taylor, 1960, p. 412, pl. 49, fig. 4.

ORDER GIGARTINALES

Plants crust-forming, foliaceous, or bushy-branched, of uniaxial or multi-axial structure and corticated. Tetrasporangia scattered on the surface or, in some species, restricted to certain areas. Spermatangia borne on surface cells in limited areas. Carpogonial filaments sunken in the cortex. Auxiliary cells differentiated before fertilization from intercalary cells of the cortex. Carpogonium producing ooblast filaments after fertilization that transmit zygote nuclei to the auxiliaries, from which goni-moblasts arise that bear carpospores.

FAMILY SOLIERIACEAE

Plants consisting of a flat sheet or of terete to flattened branches that may be sparingly branched or bushy. Medulla distinctly filamentous, the cortex pseudoparenchymatous with large inner cells and a progressive reduction in cell size outward to the small surface cells. Tetrasporangia slightly immersed, zonate in arrangement, scattered. Cystocarps large, having a central area of sterile, nurse tissue and an apical pore through which the carpospores are discharged.

GENUS *SOLIERIA*

Plants bushy, much-branched, the branches terete to somewhat flattened, and having a hollow central portion with loose, ramifying filaments and a soft polysaccharide (lambda-carrageenan) ; cortex of cells that appear parenchymatous, as the filamentous nature is obscured, the inner cells large, and the cell size progressively smaller to the surface layer. Tetrasporangia zonate, immersed but near the surface. Spermatangia in patches on young branches. Carpogonia arising from inner cortical cells and producing a long trichogyne that comes to the surface. Cysto-carps mostly immersed.

SOLIERIA TENERA (J. AGARDH) WYNNE AND TAYLOR
(Fig. 33)

Plants 10–30 cm tall, densely branched, the branches terete or a little flattened in the main axes, yellowish to deep rose-red; branchlets usually with a constricted base and a long-tapering apex. Sporangia just beneath

Fig. 33. *Solieria tenera,* photograph of an herbarium specimen ($\times \frac{1}{4}$)

the surface, radial in position and zonate. Cystocarps scattered and forming dark, immersed, nodulelike structures that protrude only slightly above the surface.

The cell walls of *S. tenera* contain an abundance of a soft-gel-forming polysaccharide, lambda-carrageenan.

Abundant in Virginia waters wherever the salinity exceeds about $15^0/_{00}$ and whenever the water temperature is between 18° and 25° C. Known from Cape Cod (and from a few warm bays north of the Cape) south to the Caribbean Sea. In the southern part of its range, it is principally a winter and spring species. North of Virginia it is primarily a summer species. In Virginia it is present the year around. The potential commercial value of its cell-wall polysaccharides, or some other type of utilization, may eventually justify its cultivation.

J. Agardh, 1841, p. 18 (as *Gigartina tenera*); J. Agardh, 1848–76 (1851), p. 354 (as *Rhabdonia tenera*); Harvey, 1853, p. 121, pl. 23 (as *Solieria chordalis*); Schmitz, 1889, p. 441; Hoyt, 1917–18, p. 479, pl. 96; Taylor, 1957, p. 276, pl. 38, fig. 4, pl. 40, fig. 7, pl. 41, fig. 2, and pl. 59, fig. 9; Zaneveld and Barnes, 1965, p. 23; Edwards, 1970, p. 36, figs. 137–40 (as *Agardhiella tenera*); Wynne and Taylor, 1973, figs. 1–6; Hamm and Humm, 1976, p. 214; Humm, 1976, p. 78.

FAMILY HYPNEACEAE

Plants much-branched and bushy, the branches terete and often with spinelike branchlets. Growth is from a single apical cell that produces an axial filament and a parenchymalike cortex with larger cells at the center. Tetrasporangia produced in small, swollen branchlets, the spores zonate. Cystocarps swollen and mostly external, borne on small branches, without a pore.

GENUS *HYPNEA*

Plants bushy, the branches slender and cylindrical, solid, developing from an apical cell that gives rise to a more or less persistent axial filament surrounded by a cortex in which the cells are largest near the center and progressively smaller to the surface layer. Sporangia zonate, embedded in the outer cortex of slightly swollen, ultimate banchlets, though scattered rather than in sori. Cystocarps spherical, prominently protruding on the branchlets. Spermatangia superficial, each producing a row of four spermatia.

HYPNEA MUSCIFORMIS (WULFEN) LAMOUROUX

(Fig. 34)

Plants 8–30 cm tall, much-branched, olive green to purplish-green, the ultimate branchlets numerous, and often spinelike; some branches usually terminating in a hooked tip. Cystocarpic and tetrasporic plants often are quite different in appearance. Male plants are apparently rare in this region.

The cell-wall polysaccharide of *H. musciformis* is kappa-carrageenan, an economically valuable substance that is widely used as a stabilizer in foods, cosmetics, and other products. It does not gel in pure water but forms a thermally reversible gel in the presence of a solute. The nature and quantity of the solute determines the physical properties of the gel. Kappa-carrageenan may account for as much as 60% of the dry weight of the plant. (Humm, 1942, 1944, 1948, 1951*a*, 1951*b*, 1962; De-Loach, Wilton, Humm, and Wolf, 1946; Humm and Williams, 1948).

Common at some localities in Virginia during summer and early fall, especially along the Eastern Shore and on the breakwater at Cape

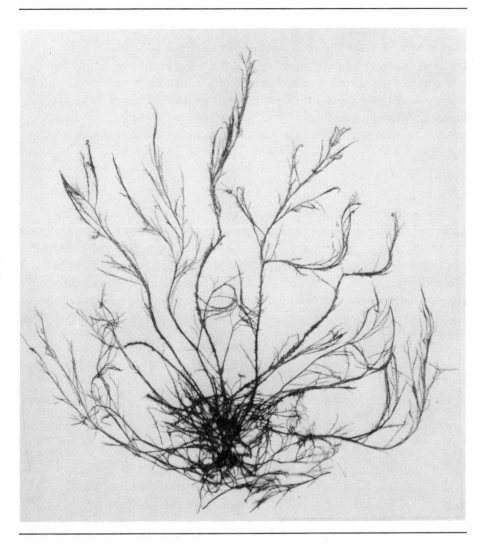

Fig. 34. *Hypnea musciformis,* photograph of an herbarium specimen (×½)

Charles. It is also present in the lower Chesapeake Bay in areas where the salinity does not fall below 20⁰/₀₀ for prolonged periods. It is present in the York River near Yorktown, for example, but perhaps not every summer. Known from the Caribbean Sea to the Gulf of St. Lawrence.

Wulfen, 1789, p. 154, pl. 14, fig. 2 (as *Fucus musciformis*); Lamouroux, 1813, p. 131; Hoyt, 1917–18, p. 485, pl. 100, and pl. 101, figs. 1–2; Taylor, 1957, p. 271, pl. 37, fig. 2; Taylor, 1960, p. 467, pl. 73, fig. 1; Kemp, 1960, p. 39; Edwards, 1970, p. 37, figs. 143–47.

FAMILY GRACILARIACEAE

Plants firm to cartilaginous, branched, sometimes bushy, the branches terete to flattened. Tetrasporangia scattered, tetrapartite, and just below the surface layer of cells. Spermatangia also scattered, produced by the surface cells. Cystocarps prominent, mostly external, with a thick pericarp having an ostiole.

GENUS *GRACILARIA*

Plants branched and bushy (in Virginia), the branches cylindrical to flattened, solid, gelatinous-cartilaginous; growth from an apical cell but not producing a distinguishable axial filament as in *Hypnea*. Medulla of large colorless cells, the size of the cells progressively reduced toward the outside to the small surface layer. Sporangia produced near the surface, scattered, the spores tetrapartite. Cystocarps large, hemispherical, prominently protruding or on the surface, the carpospores discharged through a pore.

Key to the Species of the Genus *Gracilaria*

Plants rose-red; branches terete, slender . . . . . *G. verrucosa*

Plants olive green to purplish-green; branches slightly to distinctly flattened; branching mainly dichotomous . . . . . *G. foliifera*

GRACILARIA VERRUCOSA (HUDSON) PAPENFUSS

Plants 15–45 cm tall, much-branched in an irregular manner; the branches terete, 0.5–2.0 mm in diameter, firm-gelatinous in texture, light red to dark red in color, tapering to a point. Cells large at the center (300–450 μm), as seen in cross section, and progressively smaller toward the outside to the small surface cells. Tetrasporangia scattered over the branches, numerous, just below the surface layer of cells, 20–30 μm in diameter, 30–35 μm long. Cystocarps on the surface, firm, hemispherical, numerous.

This species was the principal raw material for an agar industry in North Carolina during World War II, when there was an agar shortage (Humm, 1942, 1944, 1951a; DeLoach, Wilton, McCaskill, Humm, and Wolf, 1946). It continues to serve as an agar source in many parts of the world. It could be cultivated during the warmer months in Virginia.

In North Carolina there is a variety of this plant that is never attached, is always sterile, and is found only in certain protected bays where wave action does not wash it all ashore during storms and where tidal currents do not move it all out. This variety persists through the winter in a depauperate vegetative condition, whereas the variety that is normally attached and reproduces sexually survives the winter (in North Carolina and Virginia, at least) either in the form of dormant holdfasts or of viable spores (Humm, 1951b; Causey et al., 1946).

G. verrucosa (previously widely known by the synonym, G. confervoides) is abundant on oyster reefs near Chincoteague, at many localities along the Eastern Shore, at Lynnhaven Inlet, and in many places in the lower Chesapeake Bay during summer and fall. Katrine deWitt has contributed many specimens from the Eastern Shore and Lynnhaven Inlet collected during the 1940s.

Known from the Caribbean Sea to Prince Edward Island. North of Cape Cod it is restricted to certain warm bays. From North Carolina northward it is a summer species.

Hudson, 1762, p. 470 (as *Fucus verrucosus*); Hoyt, 1917–18, p. 483, pl. 99, fig. 1 (as *G. confervoides*); Papenfuss, 1950, p. 195; Taylor, 1957, p. 273, pl. 38, fig. 1; Taylor, 1960, p. 441, pl. 56, fig. 2; Edwards, 1970, p. 38, fig. 152.

GRACILARIA FOLIIFERA (FORSSKAL) BØRGESEN

(Fig. 35)

Plants 10–30 cm tall, the branching moderate, predominantly dichotomous, the texture firm-gelatinous, the color usually olive, olive green, or purplish-green. Plants from the open sea tend to have more red pigment. Plants vary in degree of flattening of branches from those almost completely terete to flattened branches as much as one cm wide, especially the main axes and at the dichotomies. In flattened plants, the branches are often in the plane of the flattening, and proliferous branches may arise from the margins.

Tetrasporangia are produced by subsurface cells in the upper parts of the branches, and when mature they are 20–35 μm in diameter, 30–45 μm long, the surrounding surface cells 5–10 μm in diameter. Cystocarps on the surface, dark green or blackish, large and numerous.

Cell-wall polysaccharides of *G. foliifera* are principally agar, and make up as much as 50% of the dry weight of the plant (Kim and

Fig. 35. *Gracilaria foliifera,* photograph of an herbarium specimen of the slender form with terete branches ($\times \frac{1}{2}$)

Humm, 1965). The gel strength of agar from *G. foliifera* is usually low, however, unless the plants are soaked in dilute KOH or NaOH before the agar is extracted, a procedure that apparently reduces the degree of sulfation.

    G. foliifera is abundant in Virginia waters the year around where the salinity is 15⁰/₀₀ or more, but of greatest abundance along the Eastern Shore, especially in protected bays. Tetrasporic and cystocarpic plants are common during spring and summer, and probably male plants as well. The species occurs from the Caribbean Sea to Nova Scotia and is present the year around throughout its range.

    Forsskal, 1775, p. 191 (as *Fucus foliiferus*); J. Agardh, 1842, p. 151 (as *G. multipartita*); Howe, 1920, p. 562 (as *G. lacinulata*);

Børgesen, 1932, p. 7, fig. 1; Taylor, 1957, p. 273, pl. 38, figs. 2–3, pl. 41, fig. 1, and pl. 59, fig. 7; Kim and Humm, 1965, fig. 1; Edelstein, McLachlan, and Craigie, 1967, p. 198, fig. 19; Edwards, 1970, p. 38, figs. 149–51, 153–55; Bird, McLachlan, and Grund, 1977, figs. 6–9 (showing branches of sterile, tetrasporic, female, and male plants).

FAMILY PHYLLOPHORACEAE

Plants dichotomously branched, the branches cylindrical to much flattened and straplike, firm and tough in texture. Tetrasporangia produced in nemathecial sori. Spermatangia produced on outgrowths from surface cells. Carposporangia immersed, without a definite pericarp.

GENUS *GYMNOGONGRUS*

Plants erect, repeatedly dichotomously branched, the branches cylindrical to flattened, sometimes with proliferations along the edges. Medulla of rounded cells surrounded by a cortex of firmly coherent radial rows of small cells. Sporangia in nemathecia that form swellings or cushions on the surface, the spores in tetrapartite or cruciate arrangement. Cystocarps immersed but more or less prominent on one or both sides of the branch and with one or more pores.

GYMNOGONGRUS GRIFFITHSIAE (TURNER) MARTIUS

(**Fig. 36**)

Plants in tufts from a basal disk, 2–5 cm tall, dark-purplish to blackish, usually on rocks or oyster shells just below low tide. Branches slender, tough and wiry, 0.3–0.7 mm in diameter, the tips pointed; repeatedly dichotomous or polychotomous. Sporangial nemathecia scattered, pulvinate, on the lower sides of branches, about 1 mm in diameter.

This species bears a resemblance to *Gelidium crinale* and often occurs in the same habitat. It grows on the breakwater at Cape Charles and on oyster shells at many places along the Eastern Shore. It is known from the Caribbean Sea to southern Massachusetts.

Turner, 1808–19 (1808), pl. 37 (as *Fucus griffithsiae*); Martius, 1833, p. 27; Hoyt, 1917–18, p. 477, pl. 45, fig. 3; Taylor, 1957, p. 276.

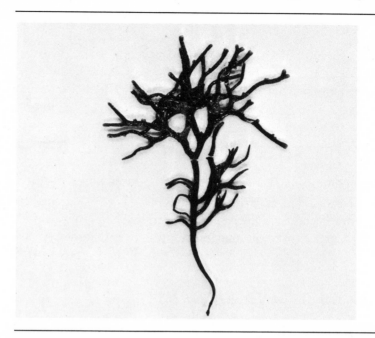

Fig. 36. *Gymnogongrus griffithsiae,* photograph of an herbarium specimen (×3)

ORDER RHODYMENIALES

Plants various in size and shape but growing from a multiaxial-type structure, with modifications, and corticated. Some plants are hollow, and many appear to be parenchymatous. Tetrasporangia in sori or scattered, just below the surface. The supporting cell of the carpogonial branch also produces one or more two-celled filaments, the terminal cell of which serves as an auxiliary cell. The carpospores are enclosed in a pericarp. There are but two families in this order, Rhodymeniaceae and Champiaceae.

FAMILY CHAMPIACEAE

Plants bushy and branched, the branches growing from a group of apical meristematic cells that produce a surface layer of small photosynthetic cells and a cortical layer of large cells around a central hollow area in which there are longitudinal filaments bearing lateral secretory cells. Sporangia produced by cortical cells and developing just below the surface. Spermatangia in sori arising from surface cells. Carpogonial filaments produced by inner cortical cells, the cystocarps immersed but conspicuous.

GENUS *LOMENTARIA*

Plants with hollow, tubular branches, terete to slightly flattened, the branching irregular. Walls of the branches thin, of three cell layers. Tetrasporangia on slightly swollen branchlets in cavities formed by a depressing of the branch wall and thus protruding into the branch cavity, scattered or in groups, the spores in tetrapartite arrangement. Spermatangia in superficial sori, borne at the ends of short filaments. Cystocarps scattered, prominent, protruding, subglobose, and with a single pore.

LOMENTARIA BAILEYANA (HARVEY) FARLOW

(**Fig. 37**, p. 94)

Plants branched, usually bushy, soft, rather gelatinous, 3–10 cm tall; the branches about 1 mm in diameter and hollow, alternate in origin, purplish-red to pink or rose. Both large and small cells are visible in surface view, the small cells often forming a reticulum. Tetrasporangia 30–50 μm in diameter, scattered, and just under the surface layer of cells. Cystocarps protruding and external on the branches, often with a slightly extended ostiolate tip.

In Virginia this species is known from the breakwater at Cape Charles and on oyster bars along the Eastern Shore in protected bays during spring and early summer. It is an annual from the Caribbean Sea to Nova Scotia. In the northern part of its range it is present in the summer, in the southern part it is present during winter and spring.

Harvey, 1853, p. 185, pl. 20 (as *Chylocladia baileyana*); Farlow, 1881, p. 154; Taylor, 1957, p. 287, pl. 35, fig. 10, pl. 41, fig. 4, and pl. 43, fig. 6; Edelstein, McLachlan, and Craigie, 1967, p. 199, fig. 37.

GENUS *CHAMPIA*

Plants similar to *Lomentaria* except that the branches are septate at intervals and constricted at each septum, with the result that the segments are somewhat barrel shaped. Sporangia tetrahedral, numerous, and scattered on the branches as dark-colored specks.

Fig. 37. *Lomentaria baileyana*, photograph of an herbarium specimen (×2)

CHAMPIA PARVULA (C. AGARDH) HARVEY
(Fig. 38)

Plants mostly 3–8 cm tall, the hollow, septate branches 0.5–1.5 mm in diameter; densely branched, terete to slightly flattened and with a beaded appearance to the unaided eye because of the barrel shaped segments. Color pink to rose, sometimes purplish to greenish. Sper-

Fig. 38. *Champia parvula,* photograph of an herbarium specimen that was an epiphyte on a *Zostera* leaf

matangia in sori in the form of bands around the branches or as caps on the ends of branches. Tetrasporangia scattered. Cystocarps ovate, scattered, sessile, and protruding.

In Virginia it is known from the breakwater at Cape Charles, where it was exceptionally abundant during the summer of 1977, from Lynnhaven Inlet (Katrine deWitt), and from many localities along the Eastern Shore.

C. Agardh, 1824, p. 207; Harvey, 1853, p. 76; Hoyt, 1917–18, p. 493, pl. 104, fig. 4; Taylor, 1957, p. 289, pl. 43, figs. 8–10; Taylor, 1960, pl. 61, fig. 4; Edwards, 1970, p. 39, figs. 162–63.

ORDER CERAMIALES

Plants composed of a main central filament that arises from an apical cell. This filament may or may not be surrounded by pericentral cells or filaments; and it may be uncorticated, corticated at the nodes only, or completely corticated. Typically the plants are much-branched, but some are not. Most are filamentous, but some are coarser and with terete branches, others flattened or strap-shaped. Carpogonial filaments are produced by an axial cell or by a pericentral cell, and the auxiliary cells are differentiated after fertilization by the supporting cell of the carpogonial branch. The cluster of carpogonia may be naked, subtended by a few short branchlets (involucre), or enclosed in a pericarp to form an ostiolate cystocarp. There are four families in the order, all of which are represented in Virginia waters.

FAMILY CERAMIACEAE

Plants of uniseriate, branched filaments that are uncorticated, corticated at the nodes only, or extensively corticated. Many produce colorless, one-celled hairs that are soon deciduous. Clusters of carposporangia may be naked, or partly protected by a few subtending, incurved filaments, the involucre.

The key to the red algae on page 57 leads to all species represented in the Virginia flora by a single taxon. If there are two or more species, the key leads to the genus only. Keys to these genera will be found following the descriptions of each.

GENUS *TRAILLIELLA*

Plants filamentous, uniseriate, with a creeping, horizontal system attached by disklike holdfasts that give rise to erect filaments that are alternately branched, these producing tetraspores in their upper ends by segmentation of the axis.

TRAILLIELLA INTRICATA (J. AGARDH) BATTERS

Plants filamentous, uniseriate, distinguished by the presence of a small gland at nearly every node; usually epiphytic and forming small, entangled tufts on the host that are 1–2 cm tall, bright rose to brownish-red, arising from basal creeping filaments about 25–40 μm in diameter, the cells mostly 1–2 diameters long; branching alternate, irregular. Tetrasporangia produced near the tips of erect filaments, single or, more often, in a series, 50–60 μm in diameter when mature. Tetrasporangia apparently have not been seen on plants from the Atlantic coast of the United States, but are known on European collections (Rosenvinge, 1923–24; Newton, 1931). Harder (1948) and others have shown that *T. intricata* is the tetrasporophyte of *Asparagopsis hamifera* (Hariot) Okamura. *Asparagopsis* has not been found in Virginia.

Known from Virginia to Newfoundland. In Virginia it has been recorded as an epiphyte of *Ruppia* leaves from Guinea Marshes near Gloucester Point, collected by Dr. F. D. Ott, May 12, 1974.

Rosenvinge, 1923–24, p. 305, figs. 213–15; Newton, 1931, p. 364, fig. 219; Taylor, 1957, p. 291, pl. 45, figs. 3–5.

GENUS *ANTITHAMNION*

Plants in tufts of uniseriate, uncorticated filaments that are often alternately branched below but repeatedly oppositely branched or with whorled branches above. Chromatophores band-shaped or rounded. Gland cells are often present. Sporangia tetrapartite on small branches or sometimes replacing a branchlet. Spermatangia in patches on branchlets. Carpogonia and cystocarps arising from the lowest cell of branchlets of the last order, the cystocarp simply a mass of carpospores upon a few-celled stalk.

ANTITHAMNION CRUCIATUM (C. AGARDH) NÄGELI

(Fig. 39)

Plants forming a soft tuft of alternately branched main filaments that are densely branched above, growing to a height of 2–5 cm. Color dull rose-red. Main filaments 50–90 μm in diameter below, the cells 90–300 μm long. Lateral branches opposite, in alternately placed pairs or in fours; ultimate branchlets in two rows or occasionally unilateral. Gland cells occasional to numerous, on the upper side of three cells near the base of the ultimate branchlets. Tetrasporangia 50–60 μm in diameter, 75–85 μm long when mature, replacing branchlets of the last order on the upper side of branches, sessile, or with a short stalk. Cystocarps 200–400 μm in diameter.

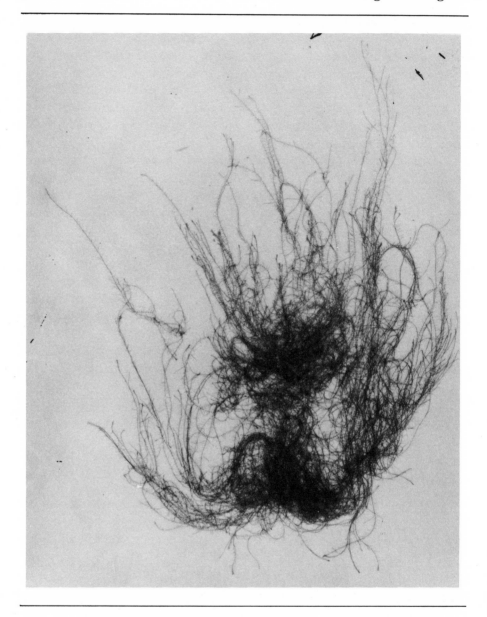

Fig. 39.   *Antithamnion cruciatum,* photograph of an herbarium specimen

On larger algae and on rocks, breakwater at Little Creek near Norfolk, July 1962; on leaves of *Zostera* at Cape Charles, October 1971.
Known from the Caribbean Sea and Bermuda to Newfoundland.

C. Agardh, 1827, p. 637; Harvey, 1853, p. 240; Farlow, 1881, p.
112 (all three as *Callithamnion cruciatum*); Nägeli, 1847, p. 200;
Rosenvinge, 1909, p. 359, figs. 294–96; Feldmann-Mazoyer, 1940, p.

254, figs. 91–94; Blomquist and Humm, 1946, p. 6; Taylor, 1957, p. 294, pl. 44, fig. 3, and pl. 45, figs. 6–8; Taylor, 1960, p. 498; Edelstein and McLachlan, 1966, p. 1054; Mathieson, Dawes, and Humm, 1969, p. 136; Whittick and Hooper, 1977, figs. 1–6; Kapraun, 1977.

GENUS *CALLITHAMNION*

Plants filamentous, uniseriate, alternately or dichotomously branched, without gland cells. Spermatangia forming small, colorless tufts near the bases of the branchlets on the upper side. Tetraspores scattered, in tetrahedral or cruciate arrangement.

Key to the Species of the Genus *Callithamnion*

1. Branching principally dichotomous . . . . *C. corymbosum*

1. Branching principally alternate 2

2. Main axes corticated by rhizoidal downgrowths from the nodes . . . . *C. baileyi*

2. Uncorticated, or only slight rhizoidal cortication 3

3. Main axes less than 50 μm in diameter; branch tips 5–7 μm in diameter . . . . *C. byssoides*

3. Main axes to 350 μm in diameter; branch tips usually 35–45 μm in diameter . . . . *C. roseum*

CALLITHAMNION CORYMBOSUM (ENG. BOT.) LYNGBYE

Plants 2–6 cm tall, rose-pink, forming dense, rounded tufts. Main branches 250–450 μm in diameter, sometimes slightly corticated by downgrowth of rhizoidal branchlets arising from the nodes, the cells 4–10 diameters long. Upper branches more or less dichotomously branched and forming corymbose tufts, the terminal cells often bearing slender, colorless hairs.

Found in Burtons Bay near Wachapreague, Virginia, by Rhodes (1970a). Previously known from New Jersey to Nova Scotia and Newfoundland on larger algae and seagrass leaves.

Lyngbye, 1819, p. 125, fig. 38C; J. Agardh, 1848–76 (1851), p. 41; Rosenvinge, 1909, p. 325, figs. 240–48.

CALLITHAMNION BAILEYI HARVEY

Plants soft, bushy, light red, tending to be pyramidal in form, arising from a much-branched filamentous holdfast. Alternately branched, the principal branches tending to be long and corticated by rhizoidal down-growths from the nodes; the ultimate branchlets curved, 20–40 μm in diameter, with cells 3–5 diameters long. Tetrasporangia scattered, sessile on the upper sides of branchlets, 70–90 μm long, 55–70 μm in diameter. Cystocarps bilobed, to about 200 μm in diameter, usually near the axil of a branch. Antheridia in rounded groups on the upper end of the lower cells of branchlets.

Collected in the York River near Gloucester Point by Dr. B. L. Wulff. Known from Virginia to Nova Scotia.

Harvey, 1853, p. 231; Farlow, 1881, p. 127; Taylor, 1957, p. 299, pl. 44, fig. 5, and pl. 46, figs. 6–9.

CALLITHAMNION BYSSOIDES ARNOTT

Plants forming soft, delicate, pink-to-rose tufts; branches of uniseriate filaments that are alternately or pinnately branched. Main axes 25–40 μm in diameter, the slender ultimate branchlets 10–20 μm in diameter, the terminal cell even less. Tetrasporangia sessile on the upper side of branchlets, obovate to subglobose, 25–40 μm in diameter, 35–50 μm long.

In Virginia it has been found in Burtons Bay near Wachapreague on the Eastern Shore, on a breakwater at Cape Charles, and near the Norfolk end of the Hampton Roads bridge-tunnel. All these collections were made in June or July. The species has been recorded from Nova Scotia to the Caribbean Sea, usually as an epiphyte.

Arnott in Hooker, 1833, p. 342; Børgesen, 1916–20, p. 218, figs. 205–7; Williams, 1948, p. 692, fig. 19; Taylor, 1957, p. 297; Edwards, 1969, figs. 1–9; Edwards, 1970, p. 39, figs. 164–68. Kapraun, 1978.

CALLITHAMNION ROSEUM (ROTH) HARVEY

Plants 2–6 cm tall, rose-pink or red, repeatedly alternately and radially branched to form a dense tuft. Main axes 125–350 μm in diameter, the upper branching with pinnate ultimate determinate branchlets, about 35–45 μm in diameter. Cells 2–4 diameters long above, 3–4 diameters long below. Tetrasporangia 60–85 μm long, 45–70 μm in diameter, often in pairs.

Reported for Virginia from Burtons Bay near Wachapreague by Rhodes (1970*a*), a winter collection; found at Cape Charles in June 1974 by Dr. F. D. Ott. Previously known from New Jersey to Cape Cod on larger algae and seagrasses.

Roth, 1797–1806 (1798), p. 46 (as *Ceramium roseum*); Rosenvinge, 1909, p. 331, figs. 249–59; Collins and Hervey, 1917, p. 136.

GENUS *SPERMOTHAMNION*

Plants uniseriate, uncorticated, branched, consisting of a basal, horizontal system of filaments attached by disk-shaped haptera and an erect system that has opposite, or sometimes alternate, branches. Carpogonia produced near the tips of lateral branches. Sporangia produced in clusters on the upper side of branchlets, on short stalks. Spermatangia in clusters borne on the upper side of branchlets, or terminal.

SPERMOTHAMNION TURNERI (MERTENS) ARESCHOUG

Plants producing tufts of erect filaments 2–5 cm tall, 20–80 μm in diameter, the cells 3–8 diameters long, arising from horizontal filaments that are 30–45 μm in diameter with cells 3–6 diameters long, attached to the host by numerous haptera. Erect filaments with branches that are usually opposite and pinnate but may be partially alternate or secund, the ultimate branchlets more slender. Tetrasporangia produced in a series on the upper sides of branchlets near the base, usually on one-celled stalks, globose, 45–65 μm in diameter when mature. Cystocarps borne near the ends of main filaments or branches, subtended by a cluster of branchlets.

Found in Virginia on leaves of *Zostera* at Cape Charles, August 8, 1974, by Dr. F. D. Ott. Known from Florida to Nova Scotia and Newfoundland as an epiphyte of larger algae, the tetrasporic plants much more common than gametophytes.

Areschoug, 1846–50, p. 113; Rosenvinge, 1923–24, p. 298, figs. 202–12 (as *S. repens, forma turneri*); Taylor, 1957, p. 302, pl. 44, fig. 1, and pl. 45, figs. 9–11.

GENUS *GRIFFITHSIA*

Plants erect, filamentous, uniseriate, alternately to pseudodichotomously branched. Cells cylindrical, barrel-shaped or obovate, multinucleate, at first bearing a whorl of trichoblasts at the upper ends that are soon deciduous. Tetrasporangia in whorls at the nodes, tetrahedral, with or without involucral cells. Spermatia in compact tufts or caplike groups on the distal ends of outer cells of fertile branches. Procarps produced on an apical cell that becomes laterally displaced by growth and division of the cell below, with the result that cystocarps appear to be lateral on the branches and are surrounded by a few large, involucral cells.

GRIFFITHSIA TENUIS C. AGARDH

Plants 2–6 cm tall, with basal creeping filaments, attached by rhizoids; erect filaments much-branched, soft and delicate, the cells cylindrical or only a little swollen at the center, the nodes slightly constricted, 120–300 μm in diameter, 3–6 diameters long. Branches arising at or below the middle of the cell from which they originate. Cells near the apex short and bearing a whorl of compact trichoblasts. Tetrasporangia verticillate, on one- to three-celled pedicels and occurring on a few to six successive nodes, 40–50 μm in diameter when mature. Spermatangial groups terminal on one- to three-celled pedicels. Cystocarps on a one-celled stalk and subtended by six to eight large, incurved, involucral cells.

Reported for Virginia by Taylor, 1957; occasional during summer and fall in the lower Chesapeake Bay and along the Eastern Shore. Known from the Caribbean Sea to Cape Cod.

G. globulifera Harvey is another species that is known from the Caribbean Sea to Prince Edward Island but apparently has not yet been found in Virginia. There is an attenuate form of this species that may be confused with G. tenuis.

Harvey, 1858, p. 130 (as Callithamnion tenue); Børgesen, 1916–20, p. 462, fig. 423; Humm and Darnell, 1959, p. 274; Edwards, 1970, p. 40, fig. 169.

GENUS CERAMIUM

Plants usually erect and bushy, though sometimes creeping; the branching usually dichotomous, but alternate in a few species. Axis a uniseriate filament with a band of corticating cells at each node or, in some species, completely corticated with small cells. Spermatangia of minute, colorless cells produced on corticating cells. Cystocarps lateral, often with a loose involucre. Tetrasporangia spherical, sessile, tetrahedral, borne at the nodes in the incompletely corticated species.

Key to the Species of the Genus *Ceramium*

1. Corticating cells completely covering the axes or nearly so . . . .
. . . . 2

1. Corticating cells covering the nodes only 3

2. Older parts with large cells over the nodes that are only partly covered by small surface cells; internodes sometimes naked at center . . . . *C. rubriforme*

2. Older parts with little difference on the surface between nodes and internodes; internodes all covered . . . *C. rubrum* (p. 104)

3. Sporangia distinctly external to the corticating cells . . . . .
. . . . *C. fastigiatum* (p. 105)

3. Sporangia mostly or completely immersed in the nodal cortication 4

4. Plants with numerous lateral adventitious branchlets; lowest corticating cells of older nodes 12–15 μm in diameter . . . . .
. . . . *C. diaphanum* (p. 105)

4. Without many lateral adventitious branches; lowest corticating cells of older nodes 17–20 μm in diameter . . . .
. . . . *C. strictum* (p. 106)

CERAMIUM RUBRIFORME KYLIN

Plants to about 10 cm tall, much-branched in an irregular fashion below but more or less dichotomously branched above, the tips distinctly forcipate and bearing colorless, deciduous hairs. Cortical cells may not completely cover the internodes in some parts of the plant, although they are usually completely covered in the older parts. One or two bands of larger corticating cells at the nodes are only partly covered by the small superficial cortical cells. Sporangia nearly immersed in the nodal cortication. Cystocarps borne on the upper branches and subtended by two short involucral branchlets.

Reported for Virginia from Burtons Bay near Wachapreague by Rhodes (1970*a*) during summer. Previously known from Massachusetts to Newfoundland.

As a result of culture work, Garbary, Grund, and McLachlan, 1978, have shown that the characters used to separate *C. rubriforme* and *C.*

rubrum vary with environmental conditions. Accordingly, they regard
C. rubriforme as a synonym of *C. rubrum.*

Kylin, 1907, p. 183, pl. 7, fig. 7; Taylor, 1957, p. 314, pl. 48, fig. 10,
pl. 49, fig. 9, pl. 50, fig. 7, and pl. 52, fig. 4; Edelstein, Craigie, and
McLachlan, 1967, p. 196, fig. 18; Mathieson, Dawes, and Humm,
1969, p. 136.

CERAMIUM RUBRUM (HUDSON) C. AGARDH

(Fig. 40)

Plants 4–15 cm tall, sometimes larger, red in color, much-branched, the
branches mostly 0.5–1.0 mm in diameter, dichotomous to somewhat
irregular, the tips forcipate or straight. Cortical cells completely covering the branches except at the tips, the nodes slightly swollen or not at
all. Tetrasporangia 50–80 μm in diameter, forming a row around the

Fig. 40.   *Ceramium rubrum,* photograph of an herbarium specimen (×1½)

node, immersed or only slightly emergent. Spermatangia in tufts near the branch tips. Cystocarps on short, lateral branchlets, subtended by an involucre of three to six branchlets that almost enwrap them.

Widely distributed in Virginia, where it is present the year around, but perhaps more abundant during winter and spring. Known from the northern Gulf of Mexico, the east coast of Florida north of Cape Canaveral to the Arctic. It is uncommon south of North Carolina, and then present only during winter and spring.

Hudson, 1762, p. 27 (as *Conferva rubra*) ; C. Agardh, 1817, p. 60; Hoyt, 1917–18, p. 514, pl. 111, fig. 2; Humm, 1952, p. 20; Taylor, 1957, p. 315, pl. 47, fig. 1, and pl. 52, figs. 5–7; Wilce, 1959, p. 73; Wulff and Webb, 1969, p. 30.

CERAMIUM FASTIGIATUM (ROTH) HARVEY

Plants forming rather dense tufts 1–3 cm tall, sometimes considerably taller, dichotomously branched, the apices usually incurved. Young internodes with only two rows of corticating cells; older nodes with three to six rows, the lowest band the largest, the cells irregular in shape but often wider than tall. Diameter of the older nodes 50–150 μm, the corticating band wider than long. Tetrasporangia one to six at a node, greatly projecting, 35–65 μm in diameter, 50–70 μm long, often oval in shape. Cystocarps single or in groups of two or three, with two to four short, involucral branchlets.

In Virginia the species has been found near Wachapreague in Burtons Bay, and it has been taken by dredge near the mouth of the Chesapeake Bay.

Roth, 1797–1806 (1800), p. 175 (as *Conferva fastigiata*) ; Harvey, 1834, p. 303; Taylor, 1957, p. 309, pl. 47, figs. 3–5, 7, pl. 48, figs. 2–4, pl. 49, figs. 3–4, pl. 50, fig. 4, and pl. 51, figs. 6–7; Taylor, 1960, p. 526, pl. 67, figs. 4–6; Zaneveld and Barnes, 1965, p. 24; Rhodes, 1970a, p. 62; Hamm and Humm, 1976, p. 214.

CERAMIUM DIAPHANUM (ROTH) HARVEY

Plants 3–10 cm tall, dichotomously branched, brownish-red in color, corticated only at the nodes, the nodes and internodes sharply distinct, the branch tips forcipate. Older nodes 300–450 μm in diameter, shorter than broad, notably wider than the internodes. Outer nodal cells small, both above and below, and partly covering the larger, central nodal cells. Tetrasporangia immersed in the upper parts of the nodes, about

50–75 μm in diameter. Cystocarps on short, lateral branches, more or less enwrapped by three or four involucral branchlets. Spermatangia produced in patches on the upper nodes.

Collected in Virginia by Michael Castagna by dredge in Wachapreague Inlet in January, 1964. Known from Virginia to Prince Edward Island.

Harvey, 1846–51, pl. 193; Taylor, 1957, p. 311, pl. 47, fig. 2, pl. 48, figs. 7–9, and pl. 51, figs. 1–4; Zaneveld and Barnes, 1965, p. 24.

CERAMIUM STRICTUM (KÜTZING) HARVEY
(**Fig. 41**, p. 107)

Plants usually 5–8 cm tall, in tufts, dichotomously branched, dull red in color, the branch tips often forcipate. Corticating cells at the nodes only, the naked internodes long and conspicuous. Nodes 200–300 μm or more in diameter, wider than the internodes, and having large cells in the center that are mostly covered in the older nodes but exposed in the younger. Internodes 150–270 μm wide in the older parts of the plants. Tetrasporangia immersed and in a whorl at the central part of the nodal band, 50–60 μm in diameter, 60–70 μm long. Cystocarps lateral on upper branches, more or less enwrapped by four to six involucral branchlets.

Because of intergrading forms and the consequent difficulty of distinguishing them, Feldmann-Mazoyer (1940, p. 302) treats *C. strictum* as a variety of *C. diaphanum*.

C. strictum is widely distributed in Virginia waters and is best developed during spring and early summer. In North Carolina it usually appears annually in February and disappears in April. It has been reported from a few places in the Caribbean Sea and is known from Florida to Nova Scotia and Prince Edward Island.

Harvey, 1846–51, p. 163; Taylor, 1957, p. 310, pl. 47, fig. 6, pl. 48, figs. 5–6, and pl. 51, fig. 5; Edwards, 1970, p. 41, figs. 176–78; Ott, 1973, p. 278.

GENUS SPYRIDIA

Plants erect and bushy, abundantly alternately branched, the branches consisting of an axial row of large cells that are completely corticated by transverse bands of longitudinally elongate cells; these branches bearing ultimate, determinate, deciduous branchlets that are corticated at

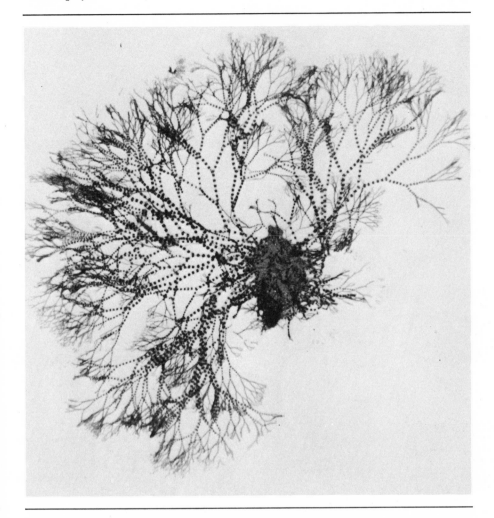

Fig. 41. *Ceramium strictum*, photograph of an herbarium specimen (×2)

the nodes only by a single ring (sometimes two) of small cells, and spine tipped.

SPYRIDIA FILAMENTOSA (WULFEN) HARVEY
(Fig. 42)

Plants 6–15 cm in height, densely branched in an irregular or alternate fashion; branches with a central, uniseriate axis of large cells, these completely corticated by alternate bands of longer and shorter surface cells, the shorter bands lying over the nodes. Ultimate branchlets determinate, corticated at the nodes only by a single row of small cells,

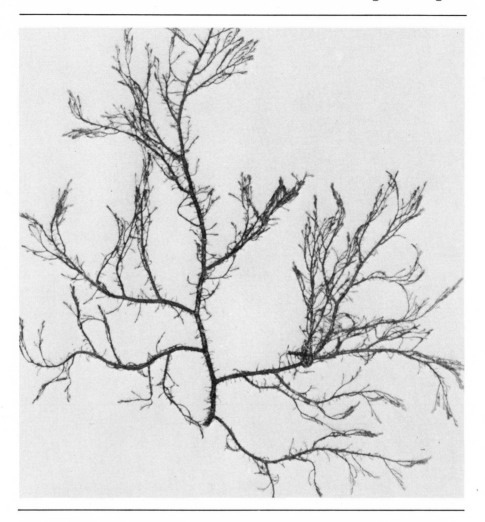

Fig. 42. *Spyridia filamentosa,* photograph of an herbarium specimen ($\times 1\frac{1}{2}$)

terminating in a spine, and finally deciduous. In the older parts of the plant, the corticating cells of the indeterminate axes become irregular in arrangement. Tetrasporangia sessile on the nodes of the branchlets, 40–70 μm in diameter. Spermatangia produced at the nodes of the branchlets and extending over the internodes. Cystocarps terminal on short branchlets, subtended by incurved involucral branchlets.

In Virginia, common on the breakwater at Cape Charles and at many other localities along the Eastern Shore during summer and fall.

Wulfen, 1803, p. 64 (as *Fucus filamentosus*); Harvey in Hooker, 1833, p. 336; Farlow, 1881, p. 140, pl. 10, fig. 1, and pl. 12, fig. 2; Hoyt, 1917–18, p. 512, pl. 111, fig. 1; Blomquist and Humm, 1946, p. 7, pl. 3, fig. 30; Taylor, 1957, p. 317, pl. 44, fig. 2, and pl. 46, figs. 2–5.

FAMILY DELESSERIACEAE

Plants of flat blades that are unbranched or branched either alternately or dichotomously; growth from a prominent apical cell that produces a primary axial row from which lateral cell rows arise, all of which are coalesced to form a flat blade that is uncorticated or corticated. Sporangia and spermatangia produced in superficial sori. Following fertilization of the carpogonium, the supporting cell of the carpogonial branch develops auxiliary cells from which the carposporophyte is produced. It becomes enclosed in an ostiolate pericarp. Cystocarps are borne along the midrib or scattered.

GENUS *CALOGLOSSA*

Plants consisting of segments of flat, linear, lanceolate, or ovate blades, dichotomously branched, developing from an apical cell and having a midrib that consists of an axial row of large cells with elongated cells on each side. Spreading from each side of the midrib is a monostromatic membrane of rows of cells that form an angle with the midrib. Proliferous branching often results in blades arising from the midrib. Tetrasporangia formed on the blade at the upper end. Cystocarps sessile on the midrib.

CALOGLOSSA LEPRIEURII (MONTAGNE) J. AGARDH

(Fig. 43)

Plants prostrate and spreading or somewhat tufted, greenish-purple, the blades 1–2 mm wide, 4–6 mm long, linear, lanceolate, sometimes ovate, narrow at the nodes, attached by rhizoides produced from the nodes.

In Virginia, it is often found around the base of the salt marsh grass *Spartina* and on old woodwork in the intertidal zone. Known from the Caribbean Sea to Connecticut from brackish water of low salinity to high salinity sea water.

Montagne, 1840, p. 196, pl. 5, fig. 1 (as *Delesseria leprieurii*); J. Agardh, 1848–76 (1876), p. 499; Taylor, 1957, p. 319, pl. 53, figs. 2–3; Taylor, 1960, p. 544, pl. 68, fig. 1; Papenfuss, 1961, figs. 1–30.

GENUS *GRINNELLIA*

Plants in the form of a flat sheet with a distinct midrib that does not extend to the upper end of the blade and without secondary veins; color

Fig. 43. *Caloglossa leprieurii,* photograph of an herbarium specimen (×3)

pink to rose-red. Growth from an apical cell, the margin one cell in thickness. Sporangia in scattered, elongate sori that project somewhat. Spermatangia in scattered sori. Cystocarps scattered on the blade, enclosed in a prominent pericarp with an apical pore, elevated.

GRINNELLIA AMERICANA (C. AGARDH) HARVEY
(Fig. 44)

Plants of large, erect, unbranched or slightly proliferous blades, pink or rose-red in color, lanceolate, ovate, or oblong in shape, 10–50 cm in length, 6–12 cm wide, arising from a short stalk. Midvein prominent, extending from the base to more than half the length of the blade. Tetrasporangial sori 0.3–1.0 mm wide, 0.5–2.0 mm long, scattered

Fig. 44. *Grinnellia americana,* photograph of an herbarium specimen (×½)

over the surface, the sporangia 50–65 μm in diameter. Cystocarps also scattered over the surface of the blade, 0.3–0.6 mm in diameter. Spermatangial plants usually small, the spermatangia in scattered or confluent sori.

Since *Grinnellia* often grows in moderately deep water, it is usually collected adrift, although it also occurs in shallow water where the salinity is consistently high.

In Virginia it has been found on the breakwater at Cape Charles and adrift at many localities in the lower Chesapeake Bay and along the Eastern Shore during spring and early summer. It is known from northern Florida and the northern Gulf of Mexico to Massachusetts. In North Carolina and South Carolina it is a plant of late winter and spring, usually disappearing in April.

C. Agardh, 1820–28, p. 173 (as *Delesseria americana*); Harvey, 1853, p. 92, pl. 21; Hoyt, 1917–18, p. 495, pl. 106, fig. 1; Taylor, 1957, p. 324, pl. 30, figs. 4–7; Taylor, 1960, p. 547, pl. 69, fig. 5; Humm, 1963b, p. 523.

FAMILY DASYACEAE

The Dasyaceae are characterized by monosiphonous, pigmented, branched filaments of determinate growth that more or less cover all the branches either radially or along two sides (dorsiventrally). These filaments are usually free, but in some genera they are united to form a network. The indeterminate branches grow from an apical cell that is not persistent, as it periodically produces a lateral tuft of filaments and is lost. A new apical cell and growing tip is then produced laterally, taking the place of the preceding apical meristem. The axial cell row becomes surrounded by a ring of five pericentral cells or filaments, and in some species these become corticated by filamentous downgrowths from the bases of branches or branchlets. Tetrasporangia are produced in special polysiphonous branchlets of determinate growth, the stichidia. Spermatangia are borne on lateral branchlets and are nonpigmented. Carpogonial filaments of four cells originate at the base of a lateral tuft of filaments from a pericentral cell. Cystocarps become enclosed in an elaborate pericarp with an ostiole.

The family is represented by a single species in Virginia waters, but it is well represented around south Florida and in the Caribbean Sea.

GENUS *DASYA*

Plants erect, radially branched, the branches terete, consisting of a central filament surrounded by five pericentral cells, uncorticated, or with a more or less dense rhizoidal cortex. Branches densely covered with monosiphonous, pseudodichotomously branched, chromatophore-

bearing filaments. Tetrasporangia in elongate, stalked stichidia. Spermatangia in clusters on lateral filaments. Cystocarps surrounded by a
pericarp with a prominent apical pore.

DASYA BAILLOUVIANA (GMELIN) MONTAGNE

(**Fig. 45,** p. 114)

Plants erect to a height of 15–45 cm, light to dark red, moderately
alternately branched, the branches covered with soft, slender, monosiphonous, pigmented filaments that are deciduous below in older plants.
Tetrasporic stichidia on a short stalk, 0.2–1.2 mm long; sporangia 40–
50 μm in diameter, the spores in tetrahedral arrangement. Spermatangial clusters linear-lanceolate and often terminating in a filament,
60–75 μm in diameter, 250–500 μm long. Cystocarps borne near the
end of short lateral branches, urn-shaped, to about 1 mm in diameter,
the apex extended into a short neck.

This is the northernmost species of the genus in the western North
Atlantic and is easily recognized.

In Virginia it is the most abundant during the spring but persists
through the summer. It is common in the lower Chesapeake Bay and
along the Eastern Shore. It occurs from the Caribbean Sea to Nova
Scotia. In the middle Atlantic states it is usually a plant of late winter
and spring, disappearing in April or May.

C. Agardh, 1824, p. 211; Harvey, 1853, p. 60 (as *D. elegans*); Farlow, 1881, p. 177, pl. 15, fig. 1; Hoyt, 1917–18, p. 508, pl. 110, fig. 2;
Taylor, 1957, p. 326, pl. 54, figs. 1–4; Edelstein, McLachlan, and
Craigie, 1967, p. 197, fig. 37 (all six as *D. pedicellata*); Dixon and Irvine, 1970, p. 480.

FAMILY RHODOMELACEAE

The Rhodomelaceae have an axial cell row that is surrounded by pericentral cell rows, the axial row derived from an apical cell and the pericentrals derived from the axial row by tangential divisions. In many
genera, this basic structure is, in turn, covered by a band or zone of
corticating cells that arise either from lateral cell divisions of the pericentrals or from filamentous downgrowths. The plants are usually much-
branched and bushy, and the apices of the branches usually have a tuft
of colorless hairs or trichoblasts that are deciduous. Tetrasporangia are
produced from internal derivatives of the pericentral cells, and the
spores are tetrahedral in arrangement. Spermatia are produced in clusters from trichoblast rudiments. Carpogonial filaments originate from

Fig. 45. *Daysa baillouviana,* photograph of an herbarium specimen ($\times\frac{1}{2}$)

basal segments of the trichoblasts that are polysiphonous and in which a pericentral cell serves as the supporting cell of the carpogonial filament. The cystocarps are enclosed by a well-developed pericarp with an ostiole.

The Rhodomelaceae may be the most highly evolved family of the red algae, and are one of the most abundant, especially in warmer waters.

GENUS *POLYSIPHONIA*

Plants erect, with or without creeping basal filaments; dichotomously or laterally branched, all branches polysiphonous. There is an axial cell row surrounded by four to many pericentral cell rows or filaments. Sporangia produced in the upper branches and originating from a pericentral cell, the spores in tetrahedral arrangement. Spermatangial clusters ovoid, to elongate and pointed, arising from the basal cell of a trichoblast rudiment. Cystocarps relatively large, often urn-shaped or globose and with a large ostiole.

Key to the Species of the Genus *Polysiphonia*

1.   Four pericentral cells 2

1.   More than four pericentral cells 3

2.   Main axes 60–100 μm in diameter; trichoblasts few or absent . . . . *P. subtilissima*

2.   Main axes 200–500 μm in diameter; trichoblasts few to abundant . . . . *P. harveyi*

3.   Six pericentral cells (rarely five or seven) . . . . *P. denudata*

3.   Sixteen (eight to twenty) pericentral cells . . . *P. nigrescens*

POLYSIPHONIA SUBTILISSIMA MONTAGNE

Plants mostly 3–8 cm tall, deep- to blackish-purple in color, soft, the erect filaments arising from a creeping base, more or less dichotomously branched below and alternately branched above, with four pericentral cells. Trichoblasts usually absent or few, so that the apical cell is exposed. Main axes 60–100 μm in diameter, the segments mostly 1.5–3.0 diameters long, sometimes longer; branchlets 35–50 μm in diameter, the segments 2–4 diameters long. Tetrasporangia in rows in the branches.

Polysiphonia subtilissima is the species that extends farthest up estuaries, often into essentially fresh water. It is widely distributed in Tidewater Virginia and is common in salt marshes, probably the year around but best developed during spring and summer. It is known from the Caribbean Sea to Nova Scotia.

Montagne, 1840, p. 109; Howe, 1920, p. 570; Wood and Palmatier, 1954; Funk, 1955, p. 136, pl. 22, figs. 3, 5; Edelstein, McLachlan, and Craigie, 1967, p. 200, figs. 23, 25; Edwards, 1970, p. 45, fig. 191.

POLYSIPHONIA HARVEYI BAILEY

Plants forming bushy, spreading tufts 3–10 cm tall, dark reddish-purple, the branching densely pinnate to irregular, usually from a distinct main axis. Ultimate branchlets 1–2 mm long, relatively rigid, and somewhat spinelike with numerous trichoblasts at the tips. Main axes 400–500 μm in diameter, with four pericentral cells throughout, uncorticated or only slightly so at the base of the plant. Branch segments from 0.5–2.0 diameters long, the branches originating in the axils of trichoblasts. Tetrasporangia 65–85 μm in diameter, causing some swelling or distortion of the branchlets. Spermatangial clusters ellipsoid, produced near the branch tips. Cystocarps ovate, 400–500 μm in diameter, and on a short stalk.

Abundant in Virginia throughout the lower Chesapeake Bay and along the Eastern Shore during summer and fall, but present the year around. Variety *olneyi* (Harvey) Collins, the plants 7–25 cm tall, has been reported for Virginia by Taylor (1957). Known from South Carolina to Nova Scotia.

Bailey, 1848, p. 38; Hoyt, 1917–18, p. 503, fig. 41, and pl. 108, fig. 4; Taylor, 1957, p. 332, pl. 56, figs. 6, 8, and pl. 58, figs. 1–5 (Var. *olneyi*).

POLYSIPHONIA DENUDATA (DILLWYN) KÜTZING

(Figs. 46, 47)

Plants mostly 4–15 cm tall, dark reddish purple to brownish or blackish, the branching mostly dichotomous and abundant, with six (rarely five or seven) pericentral cells. Main axes mostly 300–500 μm in diameter below, the upper branches 100–150 μm, the branchlets 35–50 μm diameter. Trichoblasts may be absent, sparse, or quite well developed. Cortical cells absent except for a few over the lowest axes of large plants. Tetrasporangia ovate to spherical, 60–100 μm in diameter. Spermatangial clusters lanceolate, 50–55 μm in diameter, 240–280 μm long. Cystocarps sessile or with a short stalk, subglobose, 330–540 μm in diameter.

This is the only species of *Polysiphonia* with six pericentral cells that is known to occur along the Atlantic coast of North America from North Carolina to Nova Scotia.

It is common on the breakwater at Cape Charles, Virginia, and at many places along the Eastern Shore and in the lower Chesapeake Bay during summer and fall.

Fig. 46. *Polysiphonia denudata,* photograph of an herbarium specimen

P. *denudata* has been reported for Florida and the Caribbean Sea, but these records should be reexamined, as *P. denudata* and *P. hemisphaerica* Areschoug have been confused in the past. The latter is a species of warmer waters. *P. denudata* probably occurs in the northern Gulf of Mexico and along the Florida east coast north of Cape Canaveral during winter.

Dillwyn, 1802–9, p. 85, pl. G (as *Conferva denudata*); Kützing, 1849, p. 824; Hoyt, 1917–18, p. 503; Taylor, 1957, p. 339, pl. 56, fig. 3, pl. 57, figs. 6–10, and pl. 59, fig. 1.

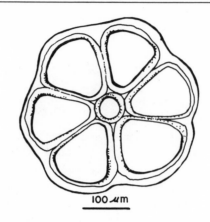

Fig. 47. *Polysiphonia denudata,* cross section of a main axis show-ing the six pericentral cells

POLYSIPHONIA NIGRESCENS (HUDSON) GREVILLE

(Fig. 48)

Plants growing to a height of 10–30 cm or more, relatively rigid in the main axes but softer above, pinnately or alternately branched from a distinct main axis, dark purple to blackish in color, and with eight to twenty, usually sixteen, pericentral cells. Tetrasporangia in the branch-lets, 60–100 μm in diameter, in rows. Spermatangial clusters lanceolate and near the tips of branches. Cystocarps broad-ovate, on a short stalk, 375–420 μm in diameter.

Common on the breakwater at Cape Charles and generally distrib-uted along the Eastern Shore and in the lower Chesapeake Bay during winter and spring, probably disappearing in May and reappearing in November or December. In North Carolina, *P. nigrescens* is usually present from December to April. It is known from South Carolina to Newfoundland.

Hudson, 1778, p. 602 (as *Conferva nigrescens*) ; Greville in Hooker, 1833, p. 322; Hoyt, 1917–18, p. 504, fig. 41, and pl. 109, fig. 3; Taylor, 1957, p. 340, pl. 56, fig. 2, pl. 58, figs. 11–12, and pl. 59, figs. 2–3; Mathieson, Dawes, and Humm, 1969, p. 138; Wulff and Webb, 1969. p. 31.

GENUS *BOSTRICHIA*

Main axes creeping, the plants dull purple, brownish, or blackish; poly-siphonous, bilaterally branched, the branches near the tips incurved, in

Fig. 48. *Polysiphonia nigrescens,* photograph of an herbarium specimen

some species with monosiphonous tips. Sporangia in stichidia in whorls, tetrahedral. Cystocarps subglobose to ovate and terminal on branchlets.

BOSTRICHIA RIVULARIS HARVEY

(Fig. 49)

Plants dull purple to greenish purple, producing some erect or partially erect branches from creeping stolons, attached by haptera. Horizontal branching dichotomous, these branches producing dorsal short branches with subcorymbose incurved branchlets that are monosiphonous only in the last one to three segments. Segments of the stolons without cortication, about half as long as broad and with six to eight pericentral cells.

Fig. 49. *Bostrichia rivularis*, photograph of an herbarium specimen (×2)

Branchlets with only four pericentral cells or fewer. Tetrasporangial
stichidia arising from the central part of the ultimate branches. Cystocarps ovate, produced on older branches.

Widely distributed along the coast of Virginia from sea water into
essentially fresh water and restricted to the intertidal zone. It occurs

around the base of salt marsh plants, especially *Spartina,* on seawalls, old woodwork, and other solid substrata. It tolerates long exposure to the air. Known from the Caribbean Sea to New Hampshire.

Harvey, 1853, p. 57, pl. 14D; Hoyt, 1917–18, p. 507; Taylor, 1957, p. 342; Taylor, 1960, p. 595.

GENUS *CHONDRIA*

Plants erect, alternately branched and bushy, the smaller branches constricted at the base and terminating in a cluster of branched, colorless hairs, and trichoblasts. Internally, the branches have an axial cell row surrounded by five pericentral cell rows and these are covered by cortical cells. Tetrasporangia embedded in the branchlets. Spermatangia in clusters that are usually flat and fan-shaped and borne near the tips of the branches on the base of a trichoblast. Cystocarps large, scattered on the upper branches, and with a well-developed, ostiolate pericarp.

There are four species of *Chondria* that are common in North Carolina and are also known to occur northward to the south side of Cape Cod (Hoyt, 1917–18; Taylor, 1957). Three of these, *C. tenuissima, C. dasyphylla,* and *C. sedifolia,* are warm-water species with tropical affinities and are primarily summer annuals from North Carolina northward. *C. baileyana,* however, is a cool-water species and flourishes during winter and spring in the southern part of its range. It is the only species of *Chondria* common in Virginia waters, and it is the most euryhaline of the four.

C. tenuissima has been tentatively recognized along the Eastern Shore by students of the Virginia Institute of Marine Science, although specimens were not preserved. *Chondria sedifolia* was reported by Rhodes (1970a) in Burtons Bay. *C. dasyphylla* is to be expected in habitats in or near the open sea along the Eastern Shore where there is suitable solid surface.

C. sedifolia and *C. dasyphylla* are so similar that they may be more properly treated as varieties of the same species. Howe (1920) regarded *C. sedifolia* as a synonym of *C. dasyphylla.* A comparison of the descriptions of these species in Taylor (1957, 1960) and Hoyt (1917–18) and the manner of their separation in the keys will serve to ilustrate the difficulty.

Key to the Species of the Genus *Chondria*

1. Main axes 0.3–1.0 mm in diameter; plants 5–10 cm tall . . . . .
    . . . . *C. baileyana*

1. Main axes 1–2 mm in diameter 2

2. Branchlet tips acute, the growing point exposed . . . . . .
    . . . . *C. tenuissima*

2. Branchlet tips blunt, the growing point sunken 3

3. Trichoblasts abundant, conspicuous; ultimate branchlets mostly 2–
    10 mm long . . . . *C. dasyphylla*

3. Trichoblasts not conspicuous; ultimate branchlets mostly 1–5 mm
    long . . . . *C. sedifolia*

CHONDRIA BAILEYANA (MONTAGNE) HARVEY
(Fig. 50)

Plants densely branched, straw-colored to greenish-purple or reddish, slender and soft, the main axes 0.3–1.0 mm in diameter, the branchlets more slender. Ultimate branchlets about 5 mm long, 80–200 μm in diameter, club-shaped, the ends a little narrowed but obtuse. Tetrasporangia in the branchlets in a band well below the apex, 50–100 μm in diameter. Cystocarps ovate, sessile on the branchlets, the tip somewhat prolonged around the ostiole.

C. baileyana is a plant of cool waters, usually appearing in Virginia in early spring and disappearing in summer. In North Carolina it appears during the winter and disappears in April. In 1962 it was growing in the Guinea Marshes near Yorktown in late June. It may persist through the warm months in the lower Chesapeake Bay and along the Eastern Shore during some years and in some localities.

Known from Nova Scotia to the Caribbean Sea, though there is reason to doubt the more southern records.

Montagne, 1849, p. 63 (as *Laurencia baileyana*); Harvey, 1853, p. 21, pl. 18; Hoyt, 1917–18, pl. 108, fig. 3 (as *C. tenuissima*, var. *baileyana*); Taylor, 1957, p. 328, pl. 55, fig. 4; Edelstein, McLachlan, and Craigie, 1967, p. 197, fig. 38.

CHONDRIA TENUISSIMA (GOODENOUGH AND WOODWARD) C. AGARDH

Plants 10–20 cm. high, tending to be tall and slender; main axes 1–2 mm in diameter. Trichoblast-bearing branchlets tapering at both ends, not crowded, the apices pointed, the trichoblasts numerous. Tetraspo-

Fig. 50. *Chondria baileyana,* photograph of an herbarium specimen

rangia in the upper half or third of the branchlets, 40–60 μm in diameter when mature. Cystocarps on short stalks, conspicuous, scattered.

Reported from Burtons Bay, Eastern Shore of Virginia, by Rhodes (1970a). Known from the Caribbean Sea to Cape Cod.

Goodenough and Woodward, 1797, p. 215, pl. 19 (as *Fucus tenuissimus*); C. Agardh, 1820–28 (1822), p. 352; Hoyt, 1917–18, p. 500; Taylor, 1957, p. 329, pl. 40, fig. 5, and pl. 55, figs. 1–3.

CHONDRIA DASYPHYLLA (WOODWARD) C. AGARDH

Plants 10–20 cm tall when mature, yellow-brown to purplish-brown; usually with a distinguishable main axis that gives rise to several similar branches, these branched in pyramidal fashion. Ultimate branchlets club-shaped, 2–10 mm long, much contracted at the base, single or in groups of two or three, the apex blunt and bearing an abundance of trichoblasts. Tetrasporangia produced in the upper part of the branchlets. Spermatangial clusters in the form of a flattened oval, arising at the base of trichoblasts. Carposporophytes enclosed by a pericarp, lateral on the branchlets, short-stalked, single or two or three in a group.

Though apparently not yet reported for Virginia, this species is known from the Caribbean Sea to Cape Cod.

Woodward, 1794, p. 239, pl. 23, figs. 1–3 (as *Fucus dasyphyllus*); C. Agardh, 1817, p. 18; C. Agardh, 1824, p. 205 (as *Laurencia dasyphylla*); Hoyt, 1917–18, p. 500, pl. 107, figs. 2, 4; Collins and Hervey, 1917, p. 121; Howe, 1920, p. 568; Edwards, 1970, p. 47, figs. 212, 214, 216.

CHONDRIA SEDIFOLIA HARVEY

Plants 8–20 cm tall, densely branched, yellow-brown, reddish-brown, or purplish-brown, the main axes 1–2 mm in diameter. Branchlets blunt at the tips, the growing point immersed in an apical pit from which arises a tuft of trichoblasts. Tetrasporangia in groups near the tips of the fertile branchlets. Cystocarps ovate, sessile or nearly so, one to three on a branchlet. Branchlets mostly 1–5 mm long.

Reported from Burtons Bay, Eastern Shore of Virginia, by Rhodes (1970a). Known from the Caribbean Sea to Cape Cod.

Harvey, 1853, p. 19, pl. 18; Hoyt, 1917–18, p. 501, pl. 108, figs. 1–2; Taylor, 1957, p. 330, pl. 55, figs. 5–6.

XANTHOPHYTA

For several decades, most phycologists have treated the plants in this group as a class of the division Chrysophyta. Because of a number of significant differences between members of the class Xanthophyceae and the other two classes traditionally placed in the Chrysophyta (Chrysophyceae and Bacillariophyceae), there has been wide acceptance of elevation of the Xanthophyceae to division status. This relatively new division has two classes, Xanthophyceae and Chloromonadophyceae. The latter are unicellular flagellates that are not well known, so the present disposition may also prove to be temporary. The Xanthophyceae contain the only benthic marine members of the division.

The species of Xanthophyceae included in this work are treated in the key to the Chlorophyta, as they look like green algae and were formerly classified with them.

CLASS XANTHOPHYCEAE

Plants of the class Xanthophyceae are yellow-green in color if they develop a relatively high proportion of carotinoid pigments or are grass green if the proportion of chlorophyll pigments is relatively high in proportion to the carotenes and xanthophylls that are always present. The plants are cellular and uninucleate or coenocytic. There are a few marine species, but the majority are restricted to fresh water.

The cells or coenocytes contain several to many chloroplasts, which are usually discoid in shape. The principal stored food is apparently chrysolaminarin, a soluble or colloidal polysaccharide; starch is not produced, but the plants may accumulate many oil droplets. Cellulose is present in the cell walls but the gelatinous polysaccharides are not well known.

ORDER VAUCHERIALES

Plants coenocytic, filamentous or expanded-saccate, with chlorophyll-bearing upper portions and rhizoidal anchoring structures that are usually colorless. Chromatophores small, disk-shaped; food stored as oil droplets. Asexual reproduction by zoospores or aplanospores. Sexual reproduction by isogametes, anisogametes, or by oogamy.

FAMILY PHYLLOSIPHONACEAE

Plants endophytic or penetrating shells of molluscs or calcareous parts of other animals; of large oval cells or filamentous; coenocytic, with many nuclei and disklike chromatophores that lack pyrenoids. Reproduction by aplanospores.

GENUS *OSTREOBIUM*

Plants of irregular, coenocytic filaments that are variable in diameter. Aplanospores formed in swollen ends.

OSTREOBIUM QUEKETTII BORNET AND FLAHAULT
(**Fig. 51**)

Plants of irregular, anastomosing filaments, mostly 4–6 μm in diameter, with local inflations of much greater diameter, the tips of the branches

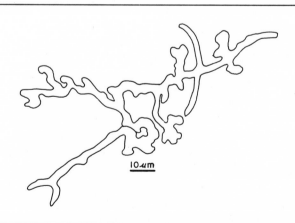

10 μm

Fig. 51. *Ostreobium queketti,* filaments from a decalcified mollusc shell

often tapering to 2 μm in diameter. Found only within mollusc shells and other animal limestone.

In old shells from Lynnhaven Inlet near Virginia Beach. Probably of general distribution along the coast of Virginia, as this is a species that is easily overlooked. Known from the Caribbean Sea to the Arctic.

Bornet and Flahault, 1889, p. 161, pl. 9, figs. 5–8; Collins, 1909, p. 408; Kylin, 1949, p. 69, fig. 66; Taylor, 1957, p. 95, pl. 2, fig. 3; Humm and Taylor, 1961, p. 373, figs. 3, 9; Edelstein and McLachlan, 1967*b*, p. 213.

FAMILY VAUCHERIACEAE

Filamentous, irregularly or dichotomously branched plants that are coenocytic; cross walls are formed, however, in connection with reproductive structures. Asexual reproduction by multiflagellate zoospores produced singly in sporangia that are terminal on branches, the flagella in pairs over the entire surface of the spore. Sexual reproduction oogamous, the plants monoecious or dioecious, the antheridia usually lateral and resembling a short, often curved, cylinder producing numerous biflagellate sperms; oogonia single or in groups, usually lateral, producing a single egg that is fertilized in place, the zygote developing a thick wall. Food is stored as oil droplets. There is but one genus, *Vaucheria*.

VAUCHERIA THURETII WORONIN
(Fig. 52)

Plants forming dense green mounds or patches in fine sand, muddy sand, or mud around the margins of salt marshes and in drainage creeks, the filaments tending to cause the accumulation of additional sediments. Emergent branches about 75 μm (30–120) in diameter, sparingly branched. Antheridia sessile, ovoid, 50–70 μm in diameter, 100–150 μm long. Oogonia sessile or on short lateral branches, obovoid to pyriform, 200 μm in diameter, 250–300 μm long; oospores subspherical, 150–180 μm in diameter. Aplanospores ovoid, 80 μm in diameter, 100–120 μm long, on short branches at right angles to the filaments. Reproductive structures produced during winter or spring.

Since only aplanospores have been found on material here recorded for Virginia, the determination is uncertain.

Along salt marsh margins of the York River northeast of the Virginia Institute of Marine Science, Gloucester Point. Known from Florida to New Brunswick.

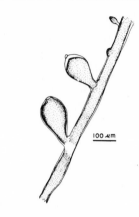

Fig. 52. *Vaucheria thuretii,* part of a filament with oogonia

Other species are surely present in Virginia waters.

Woronin, 1869, p. 157, pl. 2, figs. 30–32; Farlow, 1881, p. 104; Collins, 1909, p. 424; Hamel, 1931, p. 101, fig. 32; Humm and Taylor, 1961, p. 376.

PHAEOPHYTA

The brown algae are consistently brown or yellow-brown as a result of the predominance of the pigment fucoxanthin over the chlorophylls (*a* and *c*). Mannitol (a hexahydric alcohol), laminarin (a glucosan), and fats are the principal forms of stored food; starch is not produced. The polysaccharide cell wall constituents include algin (a straight-chain poly-mannuronide) and fucoidin (a sulfated methyl pentose). The cells are uninucleate; the nuclei have centrosomes and are often surrounded by a halo of physodes. Growth may be from an apical cell, trichothallic, or intercalary.

Life histories may involve isomorphic or heteromorphic gametophyte and sporophyte, or the gametophyte may be represented only by the gametes, produced by sporophyte plants. Gametophyte plants may be monoecious or dioecious; gametes may be isogamous, anisogamous, or oogamous. Motile cells have two laterally inserted flagella, one tinsel and one whiplash type.

The Phaeophyta are virtually all marine except for three small genera of the Ectocarpaceae that are found in fresh water connected with the sea. Some groups have reached greatest development, or are found entirely, in cold waters. This fact often leads to an underestimate of the importance of brown algae in temperate and tropical waters.

Members of the Laminariales and Fucales are highly evolved sub-aquatic plants.

The classification used here is that of Smith (1955), based upon Kylin (1933) and Papenfuss (1951).

Key to the Genera or Species of the Phaeophyta

The following key leads to the species of many of the brown algae included in this work. In the case of those genera with two or more species, however, a key to the species will be found following the genus description.

1. Plants uniseriate throughout 2

1. Plants not uniseriate throughout 11

2. Plants not more than 2–3 mm tall 3

2. Plants more than 2 mm tall 7

3. Plants 1–2 mm tall 4

3. Plants less than 0.5 mm tall, always epiphytic 5

4. Base of plant a disk; gametangia on upper parts of erect filaments
    . . . . *Hummia onusta* (p. 161)

4. Base of spreading filaments; gametangia borne near the base of
    erect filaments . . . *Ectocarpus elachistaeformis* (p. 135)

5. Part or most of plant embedded in host . . . .
    . . . . *Streblonema oligosporum* (p. 139)

5. Plants forming a disk on surface of host 6

6. With unicellular hyaline paraphyses . . . .
    . . . . *Ascocyclus magnusii* (p. 148)

6. No unicellular paraphyses . . . . . *Myrionema* (p. 147)

7. Gametangia and sporangia seriate and intercalary . . . . .
    . . . . *Pylaiella littoralis* (p. 133)

7. Gametangia and sporangia lateral 8

8. Base a hemispherical cushion giving rise to a dense cluster of erect
    filaments 0.5–1.5 cm tall . . . *Elachista fucicola* (p. 151)

8. Plants otherwise 9

9. Chromatophores ribbon-shaped . . . *Ectocarpus* (p. 134)

9. Chromatophores disk-shaped 10

10. Plurilocular gametangia in dense clusters . . . .
    . . . . *Sorocarpus micromorus* (p. 142)

10. Gametangia not in dense clusters . . . *Giffordia* (p. 140)

11. Plant a hollow sphere or vesicle, epiphytic . . . .
. . . . *Leathesia difformis* (p. 152)

11. Plant otherwise 12

12. Plant a flat crust . . . . *Ralfsia verrucosa* (p. 150)

12. Plant otherwise 13

13. Plants with air bladders 14

13. Plants without air bladders 16

14. Plants with leaflike appendages . . . *Sargassum* (p. 183)

14. Plants without leaflike appendages 15

15. Plants with flat branches and a midrib . . . *Fucus* (p. 178)

15. Branches not flattened . . . . . . *Ascophyllum* (p. 177)

16. Plants distinctly flattened 17

16. Plants not flattened 20

17. Dichotomously branched . . . *Dictyota dichotoma* (p. 145)

17. Not dichotomously branched; usually unbranched . . . 18

18. Blades 2–10 mm wide; colorless hairs forming a marginal fringe
. . . . *Desmotrichum undulatum* (p. 165)

18. Blades narrower or wider; margin without a distinct hair
fringe 19

19. Blades mostly four cells thick (to seven); surface cells not much
smaller than the inner cells . . . . . *Punctaria* (p. 168)

19. Blades more than 7 cells thick; surface cells much smaller than
inner cells . . . . *Petalonia* (p. 165)

20. Plants not branched 21

20. Plants branched 22

21. Axes 2–5 mm diameter, without constrictions, often dotted with
sori of unilocular sporangia . . . .
. . . . *Asperococcus fistulosus* (p. 172)

21. Axes 5–10 mm diameter, often with constrictions; sporangia in large patches . . . . . *Scytosiphon lomentaria* (p. 172)

22. Plants epiphytic on seagrass leaves 23

22. Plants not on seagrass leaves only 24

23. Plants mainly on eel grass; soft, slippery, densely branched; main axes about 2 mm in diameter . . . . *Cladosiphon zosterae* (p. 154)

23. Plants mainly on Ruppia; sparingly branched at top only; main axes 1 mm in diameter or less . . . *Hummia onusta* (p. 161)

24. Branching dichotomous, wide-angled; branches nodulose with clumps of small filaments . . . *Stilophora rhizodes* (p. 156)

24. Branching alternate, opposite, whorled or irregular . . . 25

25. Main axes less than 100 μm in diameter, with externally visible polysiphonous segments . . . . . . *Sphacelaria* (p. 143)

25. Main axes well over 100 μm in diameter 26

26. Main axes with transverse bands of sori; axes usually over 1 mm in diameter, somewhat hollow . . *Striaria attenuata* (p. 159)

26. No transverse bands of sori; axes less than 1 mm in diameter 27

27. Main axes with four longitudinal rows of large cells, as seen in cross section, covered by a layer of small cells . . . . *Stictyosiphon soriferus* (p. 161)

27. Main axes with many more cells internally 28

28. Axes clothed with whorls of brown filaments about 4 mm long . . . . *Arthrocladia villosa* (p. 158)

28. Axes bare except for delicate, colorless hairs . . . . *Dictyosiphon* (p. 175)

ORDER ECTOCARPALES

Plants filamentous, uniseriate, branched; cell division not localized. Gametophytes and sporophytes isomorphic. Sporophytes produce zoospores in unilocular sporangia (also known as unangia) in which meiosis usually occurs, these spores normally giving rise to gametophytes; or they may produce neutral spores, usually in plurilocular sporangia (also known as plurangia). Gametophytes produce plurilocular gametangia (plurangia); they are isogamous or anisogamous. Chloroplasts are disk-shaped or band-shaped.

Sexual reproduction in the sporophyte is the production in the unilocular sporangia of spores following meiosis; sexual reproduction in the gametophyte is the production of gametes in unilocular gametangia.

As a result of a detailed study of many representatives of what she calls the "Ectocarpus complex," Dr. Orvokki Ravanko (1970) has concluded that delimitation of all species in the Ectocarpales is not possible solely on the basis of morphological characters. Their environmentally controlled morphological plasticity is such that a number of species may represent development stages or ecophenes of other species.

FAMILY ECTOCARPACEAE

Plants filamentous, erect and free or creeping within or upon a host; filaments not coalesced into a horizontal disk.

GENUS *PYLAIELLA*

Plants filamentous, uniseriate except for occasional longitudinal walls, erect; branches opposite or irregular; plants attached by a group of rhizoidal filaments that may be partially coalesced to form a holdfast. Unilocular sporangia intercalary in a series, barrel-shaped. Plurilocular gametangia usually intercalary but sometimes terminal. Both formed through modification of vegetative cells.

PYLAIELLA LITTORALIS (LINNAEUS) KJELLMAN

Plants uniseriate, with an occasional longitudinal wall, erect, in tufts, 5–15 cm tall or more; branches opposite or irregular, 20–50 μm in diameter; chromatophores disk-shaped. Plurilocular gametangia in a chain in the lateral branches or sometimes terminal. Unilocular sporangia in the same location, usually two to thirty in a chain, 35–50 μm in diameter.

In Virginia the species was collected by Dr. Charlotte Mangum on annelid worm tubes in the York River at Sandy Point near Yorktown,

winter. It is known from Virginia northward to Labrador, Hudson Bay, and Newfoundland on larger algae and rocks.

Kjellman, 1872, p. 99; Newton, 1931, p. 124, fig. 71; Taylor, 1957, p. 102, pl. 9, figs. 1–3; Mangum, Santos, and Rhodes, 1968, p. 39.

GENUS *ECTOCARPUS*

Plants filamentous, uniseriate, usually much-branched, the branches often terminating in colorless hairs; the base consisting of rhizoids or of horizontal filaments, sometimes penetrating the substratum or host. Cells with band-shaped chromatophores that may be branched or forked. Unilocular sporangia sessile or with a short stalk, globose to ellipsoid, normally produced on the sporophyte, the spores haploid. Plurilocular gametangia of a variety of shapes, depending upon the species, producing isogametes or sometimes neutral spores.

Key to the Species of the Genus *Ectocarpus*

1. Plants 1–2 mm tall, gametangia near the base . . . .
    . . . . *E. elachistaeformis*

1. Plants larger, several cm or more when mature 2

2. Filaments 8–12 μm in diameter, densely entangled into ropelike strands by curved branchlets . . . . *E. tomentosus*

2. Filaments considerably more than 12 μm in diameter; plants not in ropelike strands 3

3. Upper branches in clustered fascicles . . . .
    . . . . *E. penicillatus* (p. 136)

3. Upper branchlets not in fascicles 4

4. Gametangia long-pointed and usually hair-tipped . . . . . .
    . . . . *E. siliculosus* (p. 137)

4. Gametangia blunt, or tapering at the tip but not ending in a hair 5

5. Gametangia fusiform, 20–40 μm wide . . . .
    . . . . *E. intermedius* (p. 137)

5. Gametangia cylindrical, 10–15 μm wide . . . .
    . . . . *E. dasycarpus* (p. 138)

ECTOCARPUS ELACHISTAEFORMIS HEYDRICH

(**Fig. 53**)

Plants virtually invisible without magnification, less than 2 mm tall, consisting of a few erect filaments 10–18 μm in diameter arising from a

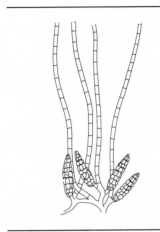

Fig. 53.   *Ectocarpus elachistaeformis*

base of wide-spreading creeping filaments, the latter 9–13 μm in diameter. Gametangia slender-conical, borne near the bases of erect filaments, 15–25 μm in diameter, 60–80 μm long or more.

On leaves of *Zostera marina* in the York River near the pier of the Virginia Institute of Marine Science, Gloucester Point. Previously known from the Gulf of Mexico and the Caribbean Sea to North Carolina.

Heydrich, 1892, p. 470, pl. 25, fig. 14; Williams, 1948, p. 686, fig. 7; Aziz and Humm, 1962, p. 61; Earle, 1969, p. 133, fig. 28.

ECTOCARPUS TOMENTOSUS (HUDSON) LYNGBYE

(**Fig. 54**)

Plants in the form of dense ropelike strands held together by numerous sharply curved branchlets, dark brown, erect, to 20 cm tall, the filaments slender, 8–12 μm in diameter; cells 2–4 diameters long with one or two band-shaped chromatophores. Plurilocular gametangia mostly on branchlets that extend out of the ropelike strands, sessile or short-stalked, ovoid to linear-obtuse, often curved, 6–18 μm in diameter, 25–100 μm

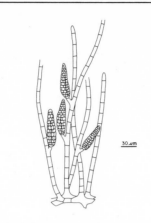

Fig. 54.   *Ectocarpus tomentosus*

long. Unilocular sporangia scattered, short-stalked, ovoid to subspherical, 15–30 μm in diameter, 20–45 μm long.

Collected in the York River near Gloucester Point, Virginia, by Dr. F. D. Ott, winter.

Hudson, 1778, p. 594 (as *Conferva tomentosa*); Lyngbye, 1819, p. 131, fig. C; Farlow, 1881, p. 70; Kylin, 1947a, p. 11, fig. 5 (as *Spongonema tomentosa*); Taylor, 1957, p. 108, pl. 8, figs. 6–8.

ECTOCARPUS PENICILLATUS (C. AGARDH) KJELLMAN

Plants to about 10 cm tall, in tufts, the upper branchlets in fasciculate groups as a result of short, curved branchlets. Gametangia sessile or short-stalked, 20–35 μm in diameter, 50–200 μm long. Unilocular sporangia sessile or short-stalked, ovate to ellipsoid, 25–30 μm in diameter, 35–50 μm long.

E. penicillatus may be simply a variety of *E. intermedius* (*E. confervoides*). It is treated by Rosenvinge (1909) as forma *penicillata* (C. Agardh) Kjellman. Kjellman (1872) originally regarded it as a form of *E. confervoides* but elevated it to species status in 1890.

From the York River near Yorktown beside the naval weapons station pier, collected by Dr. Charlotte Mangum (Mangum, Santos, and Rhodes, 1968). Previously known from Connecticut to New Hampshire.

Kjellman, 1890, p. 76; Rosenvinge, 1909, p. 34, fig. 11 (as *forma penicillata*); Taylor, 1957, p. 107.

ECTOCARPUS SILICULOSUS (DILLWYN) LYNGBYE

(Fig. 55)

Mature plants usually 15–30 cm tall, the branching often pseudodichotomous below but alternate above. Cells in the main axes 40–60 μm in

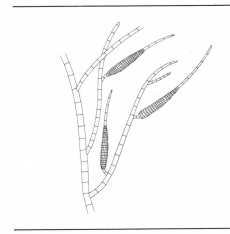

Fig. 55. *Ectocarpus siliculosus*

diameter, 4–5 diameters long; cells higher up often only one diameter long, or even shorter. Chromatophores irregularly band-shaped. Plurilocular gametangia sessile or short-stalked, conical-subulate, 50–600 μm long, 12–25 μm in diameter, often ending in a colorless hair. Unilocular sporangia sessile or short-stalked, ellipsoid, 30–65 μm long, 20–27 μm in diameter, often on the same plant with plurilocular gametangia.

Abundant along the coast of Virginia during the winter months, disappearing in April or May and reappearing in November or December. In the northern part of the Gulf of Mexico and from the north side of Cape Canaveral, Florida, to the Arctic.

Dillwyn, 1802–9, supp., p. 69, pl. E (as *Conferva siliculosa*); Lyngbye, 1819, p. 131, pl. 43 (except variety *beta* and synonyms); Hoyt, 1917–18, p. 438, fig. 8; Taylor, 1957, p. 105, pl. 8, figs. 4–5; Earle, 1969, p. 133, fig. 18.

ECTOCARPUS INTERMEDIUS KÜTZING

(Fig. 56)

Similar in vegetative characters to *E. siliculosus*, but with gametangia 60–150 μm long, 20–35 μm in diameter, short-subulate to fusiform but

Fig. 56. *Ectocarpus intermedius*

not ending in a hair tip. Unilocular sporangia sessile or short-stalked, ovoid, 20–40 μm in diameter, 35–50 μm long.

Abundant along the coast of Virginia at the same time of year as *E. siliculosus*.

LeJolis, 1863, p. 75; Farlow, 1881, p. 71; Hoyt, 1917–18, p. 439, fig. 9; Taylor, 1957, p. 106, pl. 8, figs. 1–3; Mathieson, Dawes, and Humm, 1969, p. 122 (all five as *E. confervoides* [Roth] LeJolis); Earle, 1969, p. 130, fig. 19.

ECTOCARPUS DASYCARPUS KUCKUCK

Plants 5–7 cm tall, pseudodichotomously branched with small lateral branches. Main axes 20–40 μm in diameter, the cells 2–3 diameters long, containing forked or band-shaped chromatophores. Gametangia usually terminal on few-celled lateral branches, but sometimes sessile and lateral, 10–15 μm in diameter and up to 250 μm long.

From shallow water, Smith Island, Northampton County, Virginia, collected by W. M. Willis (1973). Previously known from Tampa Bay, Florida, along the Louisiana and Mississippi coasts, and from southern Massachusetts to Rhode Island.

Kuckuck, 1891, p. 97, fig. 4; Taylor, 1957, p. 105; Humm and Caylor, 1957, p. 248; Earle, 1969, p. 132.

GENUS *STREBLONEMA*

Plants epiphytic and creeping upon, or penetrating, the host; of uniseriate, branched filaments that are not appressed into a disk; without erect external branches, although erect, colorless hairs are produced. Gametangia erect, cylindrical, sometimes branched, uniseriate or pluriseriate. Unilocular sporangia oval to spherical.

STREBLONEMA OLIGOSPORUM STRÖMFELT

(Fig. 57)

Filaments creeping within the polysaccharide layer or between the cells of the host, 5–10 μm in diameter, and bearing erect, external, colorless hairs with basal growth. Plurilocular gametangia external, usually uniseriate but sometimes biseriate in part, 8–15 μm in diameter, 25–40 μm

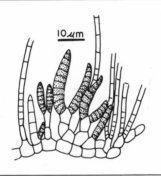

Fig. 57. *Streblonema oligosporum,* a portion of a plant in section showing gametangia and hairs with basal growth

long, sessile on horizontal filaments or terminal on filaments directed toward the periphery of the host. Unilocular sporangia apparently not reported.

On *Porphyra umbilicalis,* Cedar Island near Wachapreague, Virginia, January 1973, collected by Dr. Russell G. Rhodes, and from shallow water, Smith Island, Northampton County, Virginia, collected by W. M. Willis (1973). Previously known from North Carolina and from Maine to Nova Scotia.

Strömfelt, 1884, p. 15, pl. 1; Hamel, 1931–39, p. 70, fig. 21e (as *S. stilophorae*); Kylin, 1947a, p. 20, fig. 16 (as *Entonema oligosporum*); Taylor, 1957, p. 114, pl. 11, fig. 6; Aziz and Humm, 1962, p. 59, fig. 1 (as *E. oligosporum*); Rhodes, 1976, p. 177, figs. 5–9.

GENUS *GIFFORDIA*

Plants resembling *Ectocarpus* but having numerous disk-shaped chromatophores and exhibiting anisogamy, the plurilocular gametangia of two or three types differing in size of loculi and gametes.

In the older literature, species of *Giffordia* are found in the genus *Ectocarpus*.

Key to the Species of the Genus *Giffordia*

Gametangia 15–20 μm in diameter, to 150 μm long, main axes 35–45 μm in diameter . . . . *G. mitchelliae*

Gametangia 20–30 μm in diameter, to 250 μm long, main axes 20–30 μm in diameter . . . . *G. indica*

GIFFORDIA MITCHELLIAE (HARVEY) HAMEL
(**Fig. 58**)

Plants bushy, much-branched, attached at the base by rhizoids, 2–30 cm or more in height. Main axes 35–50 μm in diameter, the branches tapering to 15–20 μm in diameter and often with hairlike tips. Chromatophores discoid, numerous. Plurilocular gametangia elliptic-oblong

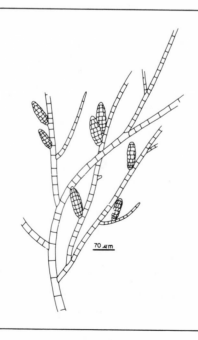

70 μm

Fig. 58. *Giffordia mitchelliae*

to long-cylindrical, blunt at the apex, 15–30 μm in diameter, 50–150 μm long, polymorphic, the smaller with cells 6–7 μm in diameter, the larger with cells 10–17 μm in diameter. Unilocular sporangia rare, oblong, 25–45 μm in diameter, 60–80 μm long.

Common along the Eastern Shore of Virginia and in the Chesapeake Bay during spring and summer. Known from Brazil through the Caribbean Sea to Nova Scotia.

Harvey, 1851, p. 142, pl. 12, fig. g; Hoyt, 1917–18, p. 439, fig. 10 (both as *Ectocarpus mitchelliae*); Hamel 1931–39, p. xiv, figs. 61 c, d; Taylor, 1960, p. 206, pl. 29, figs. 1–2; Earle, 1969, p. 138, fig. 24.

GIFFORDIA INDICA (SONDER) PAPENFUSS AND CHIHARA

Plants 2–5 cm tall when mature, arising from a rhizoidal base, irregularly and densely branched, the main branches 20–30 μm in diameter. Cells 0.5–1.5 diameters long, with discoid chromatophores. Gametangia sessile or short-stalked, usually cylindrical, and with a blunt or rounded apex, 20–30 μm (to 50 μm?) in diameter, 100–250 μm long. Unilocular sporangia sessile, oval, about 70 μm in diameter when mature and to 110 μm long.

Known from the Gulf of Mexico, the Caribbean Sea, and from Florida to Maryland. Apparently not yet recorded for Virginia, but its known distribution indicates that it is present.

Sonder in Zollinger, 1854, p. 3 (as *Ectocarpus indicus*); Taylor, 1960, p. 207, pl. 29, fig. 10 (as *G. duchassaingiana* [Grunow] Taylor); Papenfuss, 1968, p. 30; Earle, 1969, p. 136, fig. 27; Mueller, 1976, p. 78.

GENUS PORTERINEMA

Porterinema fluviatile (Porter) Waern is a microscopic member of the Ectocarpaceae characteristic of brackish water. It was reported for Virginia by Rhodes (1972), but is not included here because of the difficulty of recognizing it, except in culture.

GENUS SOROCARPUS

Plants erect, uniseriate, much-branched, the branches terminating in a colorless hair and producing hairs laterally. Gametangia in densely crowded racemose clusters at the bases of branchlets or hairs, anisogamous. Unilocular sporangia unknown.

SOROCARPUS MICROMORUS (BORY) SILVA

Plants to 20 cm tall, but usually much less, arising from a disk-shaped holdfast consisting of small cells in radial rows. Erect filaments 20–50 μm in diameter, the cells in the upper parts of the branches tending to be barrel-shaped and isodiametric, but cylindrical below and 2–5 diameters long. Chromatophores numerous and disk-shaped. All branches terminate in a colorless hair with basal growth, and hairs are also produced laterally on the branches. Plurilocular sporangia occur in dense clusters at the ends of short branches or laterally, usually at the base of a hair. Individual sporangia are ovate, 10.5–12.0 μm in diameter, 18–25 μm long.

   This species was found in Virginia at Cedar Island in March 1973 by Dr. R. G. Rhodes. It was previously known from North Carolina and from Buzzards Bay on the south side of Cape Cod, Massachusetts, to Greenland. Apparently it is uncommon throughout its range.

   S. micromorus was originally described by Lyngbye (1819, p. 132, fig. 43) as variety *uvaeformis* of *Ectocarpus siliculosus*. Pringsheim (1862, p. 9, pl. 3) named it *Sorocarpus uvaeformis* after study of a collection he obtained at Heligoland. Bory, however, transferred this plant to *Botrytella micromorus* in 1822, a name that was overlooked by most phycologists. Silva (1950) proposed *Sorocarpus* for conservation over *Botrytella,* as the former had been used so extensively, hence the name *S. micromorus.*

   Newton, 1931, p. 130, fig. 77; Hamel, 1931–39 (1931), p. 64, fig. 19; Rosenvinge and Lund, 1941, p. 58, fig. 30; Williams, 1948, p. 688, fig. 8 (all as *S. uvaeformis*); Taylor, 1957, p. 116, pl. 9, fig. 6; Pedersen, 1974, figs. 1–3; Rhodes, 1976, p. 177, figs. 1–4.

ORDER SPHACELARIALES

Plants filamentous, polysiphonous, growing from a prominent apical cell, derivatives of which divide longitudinally in a pattern characteristic of the order. Gametophyte and sporophyte similar in appearance.

FAMILY SPHACELARIACEAE

Plants arising from a holdfast or producing horizontal creeping filaments from which erect branches arise. The erect branches may be similar to the stolons, or the stolons may bear numerous, short, determinate branchlets. Cortication, if present, is rhizoidal in form. Unilocular sporangia and plurilocular gametangia produced on branchlets.

GENUS SPHACELARIA

Plants small, usually less than one but to a few cm tall, wiry, attached by a disklike holdfast, or producing stolons to form a spreading mat. Erect axes sparingly to moderately branched, usually producing determinate branchlets and also species-characteristic propagula arising from a slender stalk. Gametangia or sporangia produced on short, lateral stalks.

Key to the Species of the Genus *Sphacelaria*

1. Erect filaments 15–45 μm in diameter . . . . . . *S. furcigera*

1. Erect filaments over 50 μm in diameter 2

2. Propagula contracted at base of stalk, a short hair terminating top of stalk . .   . . *S. cirrosa*

2. Propagula not contracted at base; top of stalk with a small button-like cell only . . . . *S. fusca*

Goodband (1971) has discussed the taxonomy of these three species.

SPHACELARIA FURCIGERA KÜTZING

Plants in tufts, arising from horizontal filaments, 0.5–3.0 cm tall, the branching irregular. Erect axes 15–45 μm in diameter, producing propagula that are equally biradiate, the arms long-cylindrical, attached by a short stalk that is attenuate at the base. Anisogamous; the microgametangia cylindrical, 25–30 μm in diameter, 45–65 μm long, with cells about 3 μm in diameter; macrogametangia 28–40 μm in diameter, 30–60 μm long, the cells 5–7 μm in diameter. Unilocular sporangia globose, 50–70 μm in diameter, on one-celled stalks.

Known from the Caribbean Sea to North Carolina and from Massachusetts. Though it apparently has not yet been recorded for Virginia, it is surely present.

Kützing, 1845–71 (1855), p. 27, pl. 90, fig. 2; Blomquist and Humm, 1946, p. 5, pl. 2, fig. 14; Earle, 1969, p. 144, fig. 31.

SPHACELARIA CIRROSA (ROTH) C. AGARDH

Plants in the form of small, rounded tufts, 0.5–2.0 cm tall, the basal filaments forming a compact disklike attachment mass. Erect filaments 40–100 μm in diameter, much-branched in an alternate, opposite, or irregular manner and producing short determinate branchlets and numerous propagula that are about 500–600 μm long, usually with three arms, the arms cylindrical or tapered and with a short hair at the end of the stalk from which they arise. The base of the stalk of the propagula is much constricted at its point of attachment. Plurilocular gametangia cylindrical, blunt at the top, 70–80 μm long, 60–65 μm in diameter. Unilocular sporangia on one-celled stalks, spherical, 75–100 μm in diameter.

Known from Virginia to Baffin Island as an epiphyte on *Fucus, Ascophyllum,* and other algae and on stones, shells, barnacles, and submerged wood. Found in Virginia by Dr. F. D. Ott on old pilings of a dock at the south end of Cedar Island, February 22, 1974.

Harvey, 1851, p. 137; Newton, 1931, p. 189, fig. 118; Hamel, 1931–39 (1938), p. 258, fig. 48; Kylin, 1947*a*, p. 29, fig. 24; Taylor, 1957, p. 121, pl. 17, figs. 1–6; Goodband, 1971, figs. 1, 2, 10.

SPHACELARIA FUSCA (HUDSON) C. AGARDH

Plants forming small, soft tufts of erect, sparingly branched filaments 1–3 cm tall, 60–80 μm in diameter, that arise from a compact disk of radiating filaments in a single layer. Propagula triradiate, on a stalk that increases in diameter upwards, the arms not constricted at the base and with a spread of about 90°. Propagula stalks with a buttonlike cell at the top and without a terminal hair. Sexual reproduction unknown.

Collected in Virginia by Dr. F. D. Ott on a submerged plank of the old ferry dock at Gloucester Point, August 4, 1967.

Newton, 1931, p. 189, fig. 118 (as *S. cirrosa,* variety *fusca* Holmes and Batters); Taylor, 1957, p. 120.

ORDER DICTYOTALES

Plants erect, flattened, parenchymatous; growth from a single apical cell or from a marginal row. Sporophytes produce sporangia with four or eight large, naked aplanospores. Gametophytes oogamous. There is only one family, Dictyotaceae, with about twenty genera and one hundred species.

FAMILY DICTYOTACEAE

Plants mostly medium to large, flattened, branched or unbranched, the dichotomously branching species growing from single apical cells at the branch tips, the irregularly branched or unbranched species growing from a marginal row of apical cells. Gametophyte and sporophyte isomorphic, the gametophytes dioecious. Egg cells borne singly in oogonia forming sori on the surface; sperm cells produced in large numbers in sori of plurilocular antheridia on the surface. Gametes produced periodically in *Dictyota* and perhaps other genera, but not all. Sporangia, in which meiosis occurs, scattered on sporophyte plants, each sporangium producing four or eight nonmotile spores. Sporangia are produced continually.

GENUS *DICTYOTA*

Plants erect, mostly dichotomously branched, the branches growing from a single apical cell. Structurally, the plants are of three cell layers, a medulla of large cells containing few chromatophores, and a cortical or surface layer on each side containing many chromatophores.

DICTYOTA DICHOTOMA (HUDSON) LAMOUROUX

(**Fig. 59**, p. 146)

Plants dichotomously branched, except for proliferous branches arising along the margins in older plants, 10–30 cm tall, the branches 2–5 mm wide. Sporophyte plants produce spores constantly, but gametophyte reproduction is periodic, the gametes released during spring tides of each full moon at Beaufort, North Carolina (Hoyt, 1907, 1927), and perhaps in Virginia also. At Naples, Italy, and also the south coast of England, gametes are released during spring tides of both new and full moon, or fortnightly (Williams, 1905; Lewis, 1910).

Burtons Bay near Wachapreague on the Eastern Shore was the locality of the discovery of *Dictyota* in Virginia (Humm, 1963a). It has since been found at other localities (Zaneveld and Willis, 1976). Known from Brazil through the Caribbean Sea to Virginia.

Lamouroux, 1809, p. 42; Hoyt, 1917–18, p. 461, pl. 44, figs. 1–3; Humm and Caylor, 1957, p. 248, pl. 8, fig. 1; Earle, 1969, p. 157, figs. 49–51.

ORDER CHORDARIALES

Sporophyte branched, filamentous, macroscopic (though some are very small), producing unilocular sporangia only, or both unilocular and

Fig. 59. *Dictyota dichotoma,* photograph of a living specimen

plurilocular sporangia, but not in true sori. Gametophytes microscopic and filamentous, as far as is known; the gametangia plurilocular, the gametes isogamous.

FAMILY MYRIONEMATACEAE

Sporophyte plants minute, crustose, with a monostromatic or distromatic basal layer of laterally adherent, branched filaments that produce erect filaments of limited growth and sporangia, hairs, and, usually, paraphyses. Gametophytes microscopic, filamentous.

GENUS *MYRIONEMA*

Plant a minute cushion or disk, round or irregular in shape, of radiating, prostrate, branched filaments that form a monostromatic basal layer. From the basal layer are produced branched or unbranched filaments and hairs, but no paraphyses. Unilocular sporangia borne on the basal layer or lateral on the erect filaments, ovoid or pyriform.

Key to the Species of the Genus *Myrionema*

Reproductive organs borne on erect filaments . . . *M. corunnae*
Reproductive organs borne on the basal layer . . . *M. strangulans*

MYRIONEMA CORUNNAE SAUVAGEAU

Plants epiphytic, forming a small disk on the host 2–3 mm in diameter. Filaments of the disk 4.5–7.0 μm in diameter, these producing hairs about 5 μm in diameter and short, erect filaments 6–7 μm in diameter, 100–140 μm tall. Plurilocular gametangia formed from the erect filaments beginning at the center of the disk and progressing outwards. Mature gametangia with 2–4 celled stalks, filiform, uniseriate, or, with occasional longitudinal divisions, simple or sometimes branched, 5–7 μm in diameter, 25–120 μm tall. Unilocular sporangia unknown.

Recorded for Virginia by Dr. R. G. Rhodes as an epiphyte of *Fucus vesiculosus* in Hummock Channel near Wachapreague, winter. Previously known from Rhode Island to Maine.

Sauvageau, 1897, p. 237, figs. 14 A-F; Setchell and Gardner, 1925, p. 458; Newton, 1931, p. 151; Hamel, 1931–39, p. 91, figs. 16–18, 24; Taylor, 1957, p. 131; Rhodes, 1976, p. 178, figs. 10–13.

MYRIONEMA STRANGULANS GREVILLE
(Fig. 60)

Plants in the form of small disks 1–3 mm in diameter and often irregular in outline. From the disk arise crowded, erect filaments 50–100 μm (five to seven cells) tall, 6–11 μm in diameter, clavate in shape with cylindrical cells below, shorter and moniliform cells above. Among the erect filaments are multicellular hairs 8–13 μm in diameter. Unilocular sporangia sessile or on a one-celled stalk that arises from the basal layer or from the basal cell of the erect filaments, ellipsoid to obovoid, 20–35 μm in diameter, 35–65 μm long. Plurilocular gametangia produced on

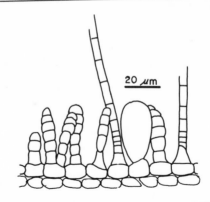

Fig. 60. *Myrionema strangulans,* a portion of a plant in section showing erect photosynthetic filaments, hairs with basal growth, and a unilocular sporangium

separate plants, obtuse-cylindrical, sessile or short-stalked, 7–11 μm in diameter, 15–50 μm long.

Found as an epiphyte of *Ulva lactuca* that was attached to a worm tube off Sandy Point in the York River near Yorktown by Dr. Charlotte Mangum. Known from Florida to Newfoundland as an epiphyte of various algae, but especially of *Ulva lactuca.*

Greville, 1823–29 (1827), pl. 300; Smith, 1944, p. 106, pl. 15, fig. 5; Williams, 1948, p. 689 (as *M. vulgare* Thuret); Taylor, 1957, p. 132, pl. 11, figs. 13–14; Mangum, Santos, and Rhodes, 1968, p. 39; Mathieson, Dawes, and Humm, 1969, p. 125; Earle, 1969, p. 179.

GENUS *ASCOCYCLUS*

Plants producing minute circular disks of radiating filaments from which arise colorless hairs, short clavate filaments, colorless unicellular paraphyses, and plurilocular gametangia.

ASCOCYCLUS MAGNUSII SAUVAGEAU

(Fig. 61)

Plants forming flat disks 1–3 mm or more in diameter, of radiating, branched, prostrate filaments from which arise erect multicellular hairs 15–20 μm in diameter, and one-celled hyaline paraphyses that are sessile

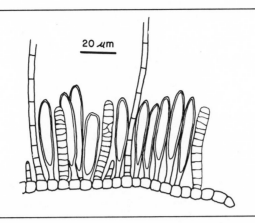

Fig. 61. *Ascocyclus magnusii,* a portion of a plant in section show-ing plurilocular gametangia, hairs with basal growth, and paraphyses

or on one-celled stalks, 8–12 μm in diameter and to 170 μm tall. Erect filaments are rare or absent. Plurilocular gametangia sessile or on one-celled stalks, cylindrical to slightly clavate, obtuse, uniseriate, 8–12 μm in diameter, 20–30 μm tall.

In Virginia this species has been collected by Dr. James Fiore at Willoughby Spit near Norfolk and by Dr. Charlotte Mangum in the York River near Yorktown. It ranges from Florida to Nova Scotia. In Florida it is apparently a winter-spring species. It is most often en-countered as an epiphyte of seagrass leaves, but it is known also from larger algae. Earle (1969) referred her collections to *A. orbicularis* (J. Agardh) Magnus, and points out (footnote, p. 178) that further study is needed to determine whether *A. magnusii* is distinct from *A. orbicularis.*

Sauvageau, 1927, p. 14; Kylin, 1947*a*, p. 40, figs. 32–33 (both as *A. orbicularis*); Edelstein and McLachlan, 1967*a*, p. 203, figs. 10–11; Mangum, Santos, and Rhodes, 1968, p. 39.

FAMILY RALFSIACEAE

Plants forming firm crusts composed of radially oriented filaments laterally united to form a flat layer from which arise a vertical layer of cells and colorless hairs. Plurilocular gametangia and unilocular sporangia formed on different plants by the upper, vertical layer, in sori, usually with paraphyses. Perennial.

GENUS *RALFSIA*

Plants usually 1–10 cm in diameter, consisting of two layers, the lower of radiating filaments and rhizoids that attach the plant, the upper of vertical ascending filaments that are photosynthetic, compact, laterally united. Unilocular sporangia and plurilocular gametangia produced on separate plants, the gametangia terminal on erect filaments, the sporangia lateral from the base of paraphyses.

Five species of *Ralfsia* are known from Connecticut or Massachusetts northward, but only one of these has been recorded to date from Virginia. Another species is known from Florida southward into the Caribbean Sea. It is possible that more than one species occurs in Virginia.

RALFSIA VERRUCOSA (ARESCHOUG) J. AGARDH

Plants forming smooth, rounded, coriaceous, brown crusts with the margins concentrically zonate, attached at first but later becoming loose in part, 0.5–10 cm in diameter, 1–2 mm thick. Filaments of the upper, erect layer 5.5–9.5 μm in diameter. Sori raised, consisting of short-stalked paraphyses, 90–170 μm tall, 4.0–5.5 μm in diameter at the base, the lowest cell several diameters long and 9–11 μm in diameter at the top. Unilocular sporangia ovoid to pyriform, 20–40 μm in diameter, 50–120 μm long. Plurilocular gametangia crowded, of one or two vertical series of cells, 7–8 μm in diameter, with the uppermost cells sterile and without paraphyses.

Recorded for Virginia by Dr. R. G. Rhodes around the base of *Spartina* plants in a salt marsh at Cedar Island, Eastern Shore, near Wachapreague. Previously known from Connecticut to the Arctic, often in tide pools or on intertidal rocks or shells, sometimes epiphytic on large algae.

Areschoug, 1843, p. 264, pl. 9, figs. 5–6 (as *Cruoria verrucosa*); J. Agardh, 1848–76 (1848), p. 62; Reinke, 1889–92 (1889), pls. 5–6; Hamel, 1931–39 (1935), p. 106, fig. 26 A-B; Taylor, 1957, p. 135, pl. 11, figs. 1–2; Rhodes, 1976, p. 179, figs. 14–17.

FAMILY ELACHISTACEAE

Plants in small tufts, consisting of a basal region of densely interwoven filaments from which erect filaments arise that may be unbranched or have short lateral branchlets at the base upon which gametangia or sporangia are borne. In some genera the reproductive structures are borne directly upon the erect branches.

GENUS *ELACHISTA*

Plants small, epiphytic, the basal part of colorless filaments, some of which may penetrate the host tissue; the basal filaments giving rise to erect, photosynthetic filaments that are branched at the base only. Interspersed among the erect filaments are paraphyses among which the sporangia are produced.

ELACHISTA FUCICOLA (VELLEY) ARESCHOUG
(Fig. 62)

Plants forming dense tufts arising from a basal, hemispherical cushion, part of which penetrates the host tissue. Erect filaments 1.0–1.5 cm tall,

Fig. 62. *Elachista fucicola,* plants growing on *Fucus vesiculosus*

densely branched below, many of the branches terminating in paraphyses or sporangia. The long, erect filaments are broader above, somewhat moniliform, and tapered at the base; about 40–50 μm in diameter at the upper ends. Unilocular sporangia ovate to pyriform, 30–90 μm in diameter, 60–100 μm long.

In Virginia found as an epiphyte of *Zostera marina* at Cape Charles,

January 1975, by Dr. F. D. Ott. Previously known from New Jersey to the Arctic, usually as an epiphyte of *Fucus* and *Ascophyllum*.

Areschoug, 1846–50, p. 377, pl. 9, fig. c; Harvey, 1851, p. 131, pl. 11, figs. 1–5; Taylor, 1957, p. 140, pl. 10, figs. 1–3; Koeman and Cortel-Breeman, 1976, figs. 1–22.

FAMILY CORYNOPHLAEACEAE

Plants forming solid cushions or spheres that are solid at first, later becoming hollow; the central tissue colorless, but this bearing paraphyses in which the chromatophores occur and form a photosynthetic cortex.

GENUS *LEATHESIA*

Sporophytes forming globose masses that are solid at first, later becoming hollow and convolute; the vaguely filamentous medulla bearing a cortex of unbranched, moniliform, photosynthetic filaments. Unilocular sporangia pyriform or ovoid, borne on the bases of the photosynthetic filaments. Plurilocular sporangia uniseriate, borne in the same position.

LEATHESIA DIFFORMIS (L.) ARESCHOUG

(Fig. 63)

Sporophyte plants 1–5 cm in diameter, sometimes more; epiphytic; at first smooth, later becoming convolute and often gas-filled; yellow-brown to olivaceous in color. Unilocular sporangia oval or oblong, about 25 μm in diameter, 35 μm long. Plurilocular sporangia uniseriate, obtuse-cylindrical, of five to ten cells, 10–16 μm in diameter, 25–35 μm long.

The plurilocular reproductive structures on sporophyte plants are believed to produce neutral (2n) spores. Gametophyte plants produce plurilocular gametangia, but the life history of this plant is still apparently poorly known.

Found as an epiphyte on *Gracilaria foliifera* on the breakwater at Cape Charles and at Chincoteague Inlet, Virginia, spring. Probably widely distributed along the Eastern Shore and in the lower Chesapeake Bay during winter and spring. Known from North Carolina to Newfoundland.

Areschoug, 1846–50, p. 376, pl. 9, fig. 8; Farlow, 1881, p. 82, pl. 5, fig. 1; Hoyt, 1917–18, p. 447, pl. 88, figs. 1–2; Taylor, 1957, p. 149, pl. 12, fig. 5, and pl. 14, fig. 8.

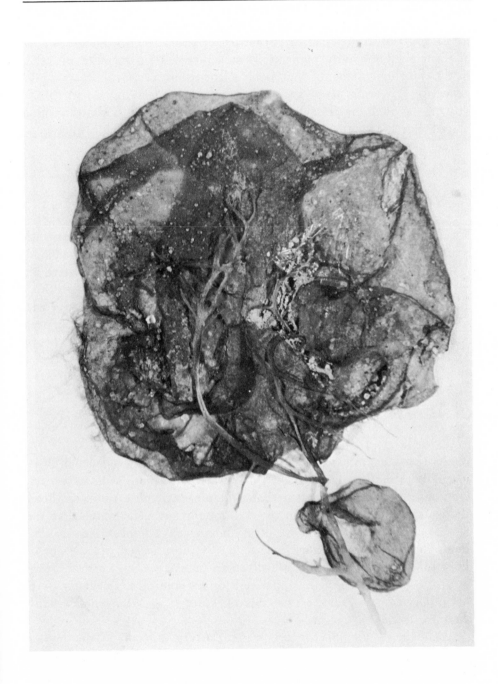

Fig. 63. *Leathesia difformis,* a large and a small plant growing on *Gracilaria foliifera.* Photograph of an herbarium specimen (×2).

FAMILY CHORDARIACEAE

Plants filiform to subspherical, branched or unbranched, sometimes hollow. Of fundamentally filamentous structure with a medulla of relatively colorless filaments that bear a cortex of erect, photosynthetic filaments. There is often an abundance of colorless hairs with basal growth. Growth of branches apical or trichothallic. Unilocular and plurilocular sporangia arising from the base of photosynthetic filaments. Gametophytes small, filamentous.

GENUS *CLADOSIPHON*

Plants soft and gelatinous, of cylindrical branches having an axis of branched, longitudinal filaments, both coarse and fine intermixed and rather closely united laterally. The longitudinal filaments give rise to an external cortex of radiating, sparingly branched, photosynthetic filaments, the whole embedded in a soft, gelatinous matrix. Plurilocular sporangia formed by lateral outgrowth of the upper cells of the cortical filaments. Unilocular sporangia borne laterally on the cortical filaments near the base.

CLADOSIPHON ZOSTERAE (J. AGARDH) KYLIN
(Fig. 64)

Plants 10–15 cm tall, soft, gelatinous, slippery, attached by a small disk, sparingly to moderately branched. Main axes 1–2 mm in diameter, the branches often bent or contorted and nodulose. Cortical photosynthetic filaments cylindrical below but of oval, moniliform cells above that are 20–40 μm in diameter and often asymmetrical. Unilocular sporangia on the base of cortical filaments, ovoid to subspherical, 35–45 μm in diameter, 55–70 μm long. Plurilocular sporangia in vertical rows formed by outgrowth of the upper cells of the cortical filaments, 7.5–18 μm in diameter and one to two cells thick, 15–45 μm long and usually three to six cells in length.

   Abundant in Virginia on eel grass, *Zostera marina,* in the higher salinity areas of the Chesapeake Bay and along the Eastern Shore, appearing each November or December and disappearing in April or May.

   The identity of this plant and related genera and species has been a source of much confusion, and as a result it has been known by a number of names now known to be synonyms, some of them widely used. Hoyt

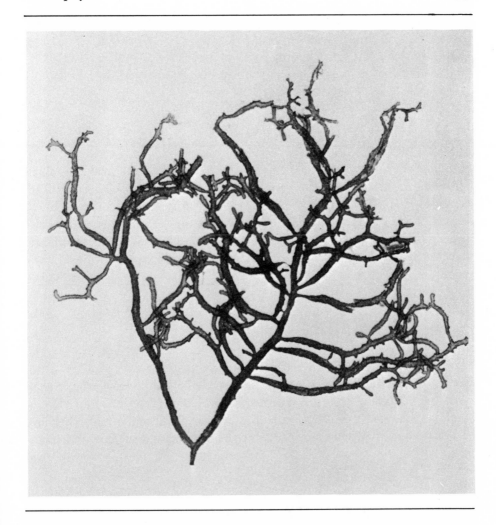

Fig. 64. *Cladosiphon zosterae*. Photograph of an herbarium specimen.

(1917–18, p. 446) comments on this problem. There is much to be learned about the life history of the plant. For example, where do the spores come from in November or December that give rise to the plants that develop on *Zostera* leaves? These *Zostera* leaves were not present the previous spring before the *Cladosiphon* plants disappeared. These circumstances suggest that there is a crop of gametophytes, or some form of the plant, present during summer or fall that provide "seed" material for the new November–December plants. If so, the warm-season form of the plant has not yet been recognized.

J. Agardh, 1841 (as *Myriocladia zosterae*); Hoyt, 1917–18, p. 446, pl. 87, fig. 1 (as *Castagnea zosterae* [Mohr] Thuret); Taylor, 1957,

p. 146, pl. 10, figs. 10–11, and pl. 12, fig. 2 (as *Eudesme zosterae* [J. Agardh] Kylin); Kylin, 1947a, p. 57, figs. 49–50, and pl. 4, fig. 15; Mathieson, Dawes, and Humm, 1969, p. 125; Earle, 1969, p. 182.

The plant has also been known as *Aegira zosterae* (Mohr) Fries.

FAMILY STILOPHORACEAE

Plants of the sporophyte bushy and branched, terete, the branches slender to moderately thick, the growth trichothallic, and giving rise to an axial group of filaments that produce a parenchymatous cortex. Colorless hairs are produced upon the cortex and sori of unilocular or plurilocular sporangia, interspersed with short filaments. Gametophyte plants are microscopic, filamentous, and they produce plurilocular gametangia.

GENUS *STILOPHORA*

Plants alternately to subdichotomously branched, the axes slender, firm, having a medulla of elongated, colorless cells and a photosynthetic cortical layer bearing colorless hairs. Unilocular sporangia ovoid. Plurilocular sporangia cylindrical and uniseriate, produced in raised, hemispherical sori that are scattered over the surface and include numerous paraphyses.

STILOPHORA RHIZODES (EHRHART) J. AGARDH
(Fig. 65)

Plants 10–30 cm tall, the axes slender, cylindrical, dichotomously to irregularly branched, the branches tapering to fine or even threadlike tips. Growth trichothallic. Branch surfaces dotted with groups of colorless hairs that are curved and about 3.5 μm in diameter at the base, 9–13 μm in diameter at the tips, 75–85 μm long. Sporangia produced at the base of surface filaments, the unilocular sporangia clavate, 22–32 μm in diameter, 36–56 μm long. Plurilocular sporangia, usually on different plants, uniseriate, composed of four to ten cells, 9–12 μm in diameter, 30–50 μm long.

Common in Virginia, especially along the Eastern Shore, in winter and spring. Not present in summer. Previously known from Florida to Prince Edward Island.

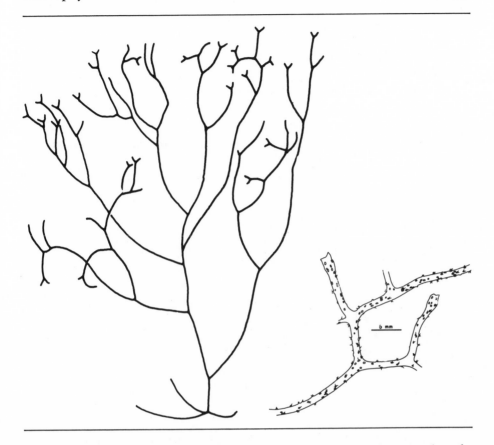

Fig. 65. *Stilophora rhizodes:* left, a habit sketch showing the dichotomous branching (×1½); right, an enlarged portion showing the raised, hemispherical sori on the surface

Ehrhart in Turner, 1808–19 (1819), vol. 4, p. 91 (as *Fucus rhizodes*); J. Agardh, 1848–76 (1848), p. 85; Hoyt, 1917–18, p. 448, pl. 87, fig. 2; Newton, 1931, p. 162, fig. 102; Taylor, 1957, p. 151, pl. 13, fig. 6, and pl. 14, fig. 6; Edelstein and McLachlan, 1967a, p. 209, fig. 35; Earle, 1969, p. 185, figs. 74, 80–82, 102.

ORDER DESMARESTIALES

Plants with trichothallic growth and of basically filamentous structure characterized by a uniaxial center with overlying cortical filaments. Sporophytes mostly of large size, producing unilocular sporangia, the spores from which give rise to microscopic gametophytes that are oogamous.

FAMILY ARTHROCLADIACEAE

Plants attached by a basal disk and having a prominent main axis and terete branches that are alternate or opposite. Growth trichothallic, producing an axial cell row of large cells covered by cortical filaments that become pseudoparenchymatous; uniseriate chromatophore-bearing filaments are produced on the surface of the branches in whorled tufts. Unilocular sporangia arising from near the base of the cortical filaments.

GENUS *ARTHROCLADIA*

Plants repeatedly branched, bushy, the main branches alternate or opposite, cylindrical, all branches bearing whorls of branched or unbranched chromatophore-bearing filaments. Internally there is a central axis of large, thick-walled cells surrounded by a cortex of several layers of thin-walled cells that are progressively smaller outward. The surface layer of cells have thicker walls and are heavily pigmented. Unilocular sporangia on short stalks, uniseriate, several to many borne on the lower part of filaments in the whorled tufts, and resembling plurilocular gametangia.

ARTHROCLADIA VILLOSA (HUDSON) DUBY

Plants arising from a basal disk to a height of 40 cm or more, the axis to 1 mm thick, alternately or oppositely branched. Branches bearing whorled tufts of brown filaments about 4 mm long, except where they have fallen off the older parts of the main branches. Unilocular sporangia occurring as branchlets of the whorled filaments, arising near their base in series of fifteen to twenty, each series on a short stalk. Individual sporangia 5.5–8.5 μm long, 11–15 μm wide; a series 50–350 μm long.

A. *villosa* is mostly restricted to moderately deep water of the open sea, where it grows on shells and stones. It is an annual and is often found washing ashore in the late summer and fall. It is common off the coast of North Carolina and is well known from New Jersey to Massachusetts, so it surely occurs off the coast of Virginia, although it apparently has not been recorded. It is to be expected on the beaches of the Eastern Shore in the fall.

Hudson, 1778, p. 603 (as *Conferva villosa*); Newton, 1931, p. 166, fig. 104; Taylor, 1957, p. 152, pl. 13, fig. 2, and pl. 17, figs. 7–8.

ORDER DICTYOSIPHONALES

Sporophytes macroscopic, parenchymatous, but never very large and not differentiated into holdfast, stipe, and blade, as in the Laminariales. Growth usually diffuse but may be subapical or basal. Gametophytes filamentous, microscopic, the majority isogamous. Sporophytes bearing unilocular or plurilocular sporangia, or both; the latter producing neutral (2n) spores.

FAMILY STRIARIACEAE

Plants filiform or slender, uniseriate to parenchymatous; growth by intercalary cell division, the gametangia and sporangia formed by transformation of surface vegetative cells.

GENUS *STRIARIA*

Plants becoming tubular, the walls of two cell layers, the inner large, the outer small. Branching opposite or irregular. Unilocular sporangia and numerous unicellular paraphyses in groups on the branches; plurilocular organs apparently not produced.

STRIARIA ATTENUATA (C. AGARDH) GREVILLE

(**Fig. 66**, p. 160)

Plants 6–8 cm tall (sometimes more), much-branched, the branches mostly 0.5–1.0 mm or more in diameter, cylindrical, usually opposite or whorled, tapering to a fine tip and also toward the base. Surface cells in irregular longitudinal rows. Unilocular sporangia developing from surface cells in groups around the axis at regular intervals of 0.25–0.50 mm, the sporangia 50–60 μm in diameter and associated with unicellular paraphyses and hairs.

Kylin (1934) cultured spores from *S. attenuata* and decided that it is a diploid species in which meiosis does not occur in the unilocular sporangia, so that the diploid spores reproduce the sporophyte asexually. Uniseriate, branched, creeping filaments (plethysmothalli) grew from the spores, and from these, typical *Striaria* plants later arose.

Collected in Virginia by Rhodes (1970a) in Burtons Bay near Wachapreague on the Eastern Shore and at Smith Island by Willis (1973). Previously known from New York to New England. In Virginia it is a winter-spring species.

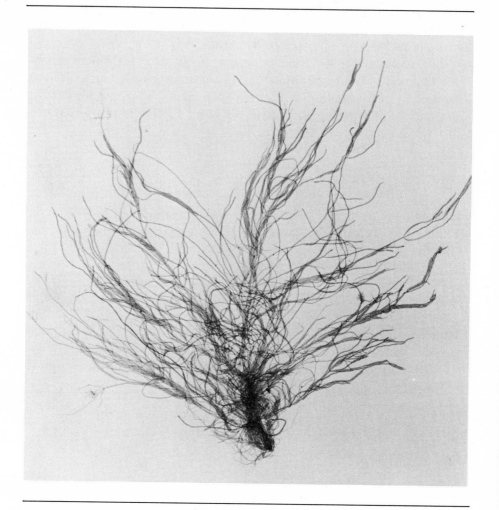

Fig. 66. *Striaria attenuata,* photograph of an herbarium specimen

Greville, 1823–29, synopsis, p. 44; Kylin, 1934, p. 15, fig. 9; Rosenvinge and Lund, 1947, p. 59, fig. 21; Taylor, 1957, p. 159; Zaneveld and Willis, 1976, p. 38.

GENUS *STICTYOSIPHON*

Plants of slender, filiform axes and branches, uniseriate to parenchymatous, the growth intercalary. Unilocular and plurilocular sporangia produced by surface cells, scattered or in sori, slightly sunken or somewhat protruding.

STICTYOSIPHON SORIFERUS (REINKE) ROSENVINGE

Plants attached by a cluster of rhizoids, the axes filamentous-poly-siphonous, to a height of 4–15 cm, and 1–2 mm in diameter. Branching irregular, the branches tapering upwards to a slender, monosiphonous tip. In the lower parts (as seen in cross section), with a medulla of four large cells surrounded by one or two layers of small cortical cells. Plurilocular sporangia (?) produced by transformation of surface cells over the polysiphonous region of the branches, or by conversion of an entire cell of the monosiphonus part. Unilocular sporangia apparently rare, but produced in the same manner as the plurilocular organs, scattered or in small groups over the polysiphonous part of the branch and somewhat protruding when mature.

Taylor (1957, p. 157) mentions this species in a footnote, as the record from New England is based upon "one fragmentary and inconclusive specimen from Massachusetts" and upon a record from Rhode Island without a specimen. Fiore (1972) found *S. soriferus* in the vicinity of Woods Hole, Massachusetts, and Rhodes (1976) found it on an oyster shell in a creek bed on Cedar Island, Accomac County, Virginia. These two records strongly suggest that the earlier records were valid.

Reinke, 1889–92 (1889), p. 59, pl. 3 (as *Kjellmania sorifera*); Rosenvinge, 1935, p. 9, figs. 9–19; Hamel, 1931–39 (1937), p. 205, fig. 42, drawings 10–14; Fiore, 1972, p. 12; Rhodes, 1976, p. 180, figs. 18–23.

GENUS *HUMMIA*

Plants with two principal morphological types in the life history: a gametophyte that is dioecious, filamentous, uniseriate, with trichothallic growth, and reproducing by plurilocular gametangia that are anisogamous; a sporophyte that is parenchymatous, filiform, with intercalary cell division, reproducing by unilocular and plurilocular sporangia that arise from surface cells of the main axes.

HUMMIA ONUSTA (KÜTZING) FIORE

(Figs. 67 and 68)

Gametophytes 1–3 mm tall, arising from a monostromatic discoid or scutate base consisting of closely appressed rows of cells. Erect axes with alternate or opposite branches, 12–15 μm in diameter, the cells

Fig. 67. *Hummia onusta,* photograph of an herbarium specimen of a group of plants of the stictyosiphon stage on a leaf of *Ruppia maritima*

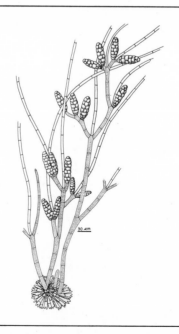

Fig. 68. *Hummia onusta,* the myriotrichia stage, showing the scutate basal disk, gametangia, and colorless hairs with basal growth

2–4 diameters long, the chromatophores disk-shaped. Colorless hairs terminal, often appearing lateral, with basal growth; growth of the filaments trichothallic and intercalary. Plurilocular gametangia of two types, borne on the upper parts of erect filaments on a short stalk, lateral, opposite, or opposite a branch, cylindrical, 18–25 μm in diameter, 30–140 μm long; female gametangia with two or three larger locules across their width, male gametangia with three to five smaller locules.

Sporophyte filiform, unbranched or with a few short branches at the top, 2–8 cm tall, 0.5–1.0 mm in diameter, epiphytic, attached by a scutate disk. Mature axis pluriseriate with four vertical rows of large cells in the center, these surrounded by one to three layers of small cells, the surface cell layer with colorless hairs that are deciduous. Unilocular sporangia 40–56 μm in diameter, 56–84 μm long; plurilocular sporangia of similar dimensions, protruding when mature.

Both gametophytes and sporophytes occur in abundance on the leaves of the seagrasses *Ruppia maritima, Zostera marina,* and *Halodule wrightii.* In Florida and the Caribbean Sea they occur to some extent on other seagrasses as well. The gametophytes apparently are year-around, at least from Virginia southward, but sporophytes in this part of the range appear each year in November and disappear in April or May. The nature of the reproduction is environmentally influenced in both gametophyte and sporophyte and exhibits seasonal rhythms, as shown by Fiore (1977) as a result of culture studies.

The gametophytes and sporophytes of this plant have been placed in different genera and species in the past, and even in different families and orders, a situation that may exist with other heteromorphic brown algae.

The gametophyte was originally described by Kützing (1849, p. 437) and illustrated in his *Tabulae Phycologicae* (1845–71, p. 22, pl. 74, fig. 1) as *Ectocarpus onustus.* Kuckuck (1956), realizing that the plant was not *Ectocarpus,* established the genus *Farlowiella* and transferred *E. onustus* to it. In the meantime, Holden (1899) had found both gametophytic and sporophytic stages of *H. onusta* near Bridgeport, Connecticut, although he thought them to be different genera. He described the sporophyte as *Stictyosiphon subsimplex* (Holden, 1899, p. 198, pl. 9, figs. a-f). The gametophyte was described by W. G. Farlow in a manuscript as *Ectocarpus subcorymbosus,* and this was later emended by Holden but not published. After Holden's death, Collins (1905) published the description.

Blomquist (1954) found the gametophyte on leaves of *Halodule wrightii* in North Carolina. He decided the plant was *Myriotrichia* and described it as a new species, *M. scutata.* Since he knew it was not an *Ectocarpus,* he did not consider all the western Atlantic species of *Ectocarpus* and did not realize that it had been named *E. subcorym-*

bosus until a few years later. He then recognized his new name for it as a synonym, and transferred it to *Myriotrichia* as *M. subcorymbosa* (Blomquist, 1958). It was still not realized that the European *Ectocarpus onustus* was the same plant.

Dr. James Fiore grew both stages in culture and discovered that the two were the same species (Fiore, 1975, 1977). He decided that there was no legitimate genus name for this plant, as *Farlowiella* is a homonym of a genus of fungi. Accordingly, he proposed the name *Hummia onusta* (Fiore, 1975). His account of its life history and reproductive behavior, based upon culture studies, was published in 1977.

The gametophyte of this plant is apparently present the year around in Virginia, the sporophyte only during winter and spring. Both are known from the Caribbean Sea to Nova Scotia.

The gametophyte: Kützing, 1849, p. 437; Kützing, 1845–71 (1855), p. 22, pl. 74, fig. 1 (as *Ectocarpus onustus*); Collins, 1905 (as *Ectocarpus subcorymbosus*); Blomquist, 1954, pl. 6, figs. 1–6 (as *Myriotrichia subcorymbosa*); Kuckuck, 1956, p. 321, fig. 14 (as *Farlowiella onusta*); Taylor, 1957, p. 109, pl. 7, figs. 5–6 (as *E. subcorymbosus*); Blomquist, 1958 (as *M. subcorymbosa*); Edelstein and McLachlan, 1967a, p. 207, figs. 12–17 (as *Farlowiella onusta*); Earle, 1969, p. 197, fig. 97 (as *Myriotrichia subcorymbosa*); Ballantine and Humm, 1975, p. 160, (as *M. subcorymbosa*); Fiore, 1975; Fiore, 1977.

The sporophyte: Holden, 1899, p. 198, pl. 9, figs. a-f; Taylor, 1957, p. 158; Edelstein and McLachlan, 1967a, p. 208, fig. 28; Cardinal, 1968, p. 258, fig. 6; Earle, 1969, p. 193, figs. 98, 105, 107; Ballantine and Humm, 1975, p. 160 (all as *Stictyosiphon subsimplex*); Fiore, 1975; Fiore, 1977, p. 303.

FAMILY PUNCTARIACEAE

Plants filiform, tubular, or of flat blades, usually attached by a basal disk, branched or unbranched, of parenchymatous structure with intercalary growth and usually with an abundance of colorless hairs. Unilocular and plurilocular sporangia produced by the transformation of vegetative cells near the surface.

GENUS *DESMOTRICHUM*

Plants filiform or ribbonlike, attached by rhizoids at the base. Lower part of the plant uniseriate or pluriseriate; the upper part pluriseriate to parenchymatous.

DESMOTRICHUM UNDULATUM (J. AGARDH) REINKE

(**Fig. 69**, p. 166)

Sporophyte plants producing light-brown linear-lanceolate blades 2–8 cm tall, 2–10 mm wide, with a disklike holdfast. Base and apex long-tapering, the blade thin-membranous, not slippery. Blade of two to four cells, 22–42 µm in thickness, the internal and surface cells alike, 10–22 µm in diameter, 7–18 µm long, usually with several small, disklike chromatophores. Marginal cells at first bearing a long hair with basal growth, 5–10 µm in diameter, but these largely deciduous. Plurilocular sporangia ovoid to conical, projecting, scattered, often marginal. Unilocular sporangia immersed, 13–22 µm in diameter, solitary or in small groups.

D. undulatum was present on *Zostera* and on various large algae in the York River near Gloucester Point, Virginia, during the winter and spring of 1961. Apparently its occurrence is irregular, but it is restricted to the winter-spring season. It was previously known from New Jersey to Nova Scotia and Prince Edward Island where it is a plant of spring and early summer.

Reinke, 1889–92, pp. 15–16, pl. 11, figs. 1–11; Farlow, 1881, p. 64 (as *Punctaria latifolia* Greville, var. *zosterae* LeJolis); Rosenvinge and Lund, 1947, p. 6, fig. 1; Taylor, 1957, p. 165, pl. 15, fig. 6, and pl. 16, figs. 1–2; Mangum, Santos, and Rhodes, 1968, p. 38, table 5.

GENUS *PETALONIA*

Plants in the form of linear-lanceolate blades with a short stalk attached by a disklike holdfast; the blade composed of large cells mixed with slender filaments and a layer of small cells on the surface with numerous chromatophores. Plurilocular sporangia develop at first in patches but later spread over the entire blade.

Key to the Species of the Genus *Petalonia*

Blade 0.6–3.0 mm wide, abruptly attenuated at the stalk . . . . . .
. . . . *P. zosterifolia*

Blade 1–5 cm or more in width, tapering at the stalk . . . *P. fascia*

Fig. 69. *Desmotrichum undulatum,* photograph of an herbarium specimen of numerous plants growing on *Gracilaria foliifera* along with other epiphytes

PETALONIA ZOSTERIFOLIA (REINKE) HAMEL

Plants narrowly linear, 7–35 cm tall, 0.6–3.0 mm wide, sometimes hollow and swollen in part, abruptly attenuating to the stalk, otherwise similar to *P. fascia*.

This plant was regarded as a narrow form of *P. fascia* by J. Agardh (1848–76, [1848]), LeJolis (1863), and others. Reinke, however, described it as *Phyllitis zosterifolia* in 1889 (1889–92). Hamel (1931–39) transferred it to *Petalonia zosterifolia*, and Rosenvinge and Lund (1941) followed this treatment, with reservations. Taylor (1957) treated it as *P. fascia*, variety *zosterifolia* (Reinke) Taylor.

P. zosterifolia represents an intermediate between the genera *Petalonia* and *Scytosiphon*. The same is true of *Scytosiphon lomentaria*, variety *complanatus* Rosenvinge, for the reason that the only difference between the two genera, as Lund has pointed out in Rosenvinge and Lund (1941, p. 37), is that *Petalonia* is flattened and solid, whereas *Scytosiphon* is terete and hollow. The difference between the two genera, hardly one of generic magnitude, is reminiscent of the difference between *Ulva* and *Enteromorpha* of the green algae, with *E. linza* representing an intermediate condition.

P. zosterifolia was reported for Virginia by Rhodes and Connell (1973). It was previously known from the north side of Cape Cod to Newfoundland.

Rosenvinge and Lund, 1941, p. 34, fig. 11; Lüning and Dring, 1973, p. 335, figs. 7–12.

PETALONIA FASCIA (MUELLER) KUNZE

Plants in the form of a blade, membranous to leathery, linear-lanceolate, 8–45 cm long, 1–5 cm in width, the tapering base often asymmetrical. Surface of blade with numerous colorless hairs; blade 145–200 μm thick where sterile, to 270 μm thick in the fertile areas. Cells in the center of the blade much larger than the outer cells (as seen in cross section), with the smallest cells at the surface. Plurilocular sporangia crowded, uniseriate, 3.5–9.5 μm in diameter, 30–75 μm long, covering broad areas in mature plants.

Ilea, another generic name for this plant, was proposed for conservation under the International Rules of Botanical Nomenclature. Silva (1952) pointed out, however, that *Ilea* is illegitimate as a genus name for this plant for several reasons and accordingly has proposed conservation of *Petalonia*.

In Virginia, *P. fascia* is abundant during winter and spring in the more saline parts of the Chesapeake Bay and along the Eastern Shore. Known from the north side of Cape Canaveral, Florida, to the Arctic, and in the Gulf of Mexico at Port Aransas, Texas (but not present in the northern Gulf of Mexico). In the northern part of its range, it is a spring and summer annual; in the southern part of its range, it appears annually about mid-November and disappears about mid-April.

Rhodes and Connell (1973) have shown that *P. fascia* passes the summer in Virginia waters in the form of a flat, brown crust that resembles the brown alga *Ralfsia*. Zoospores produced by *P. fascia* germinate to form a crust, and typical *Petalonia* blades arise from these crusts in the late fall or winter when the water temperature falls to the level to which *Petalonia* blades are adapted. There is no evidence of sexual reproduction by *P. fascia* in Virginia waters.

Wulff et al. (1968) found *Petalonia fascia* and *Scytosiphon lomentaria* common to abundant in the vegetative state on the harbor jetty at Ocean City, Maryland, during July and August 1966. These observations suggest that *P. fascia* and *S. lomentaria* are present the year around along the coast of Maryland, whereas they are present only during winter and spring along the Virginia coast. Four other species found by Wulff et al., *Nemalion multifidum*, *Polysiphonia urceolata*, *Polysiphonia novae-angliae*, and *Enteromorpha micrococca*, suggest that inshore water temperatures are significantly lower along the upper Eastern Shore of Maryland during summer, as these collections constitute the southern known limit for these species.

Kuntze, 1898, p. 419 (as *Phyllitis fascia*); Humm, 1952, p. 20; Taylor, 1957, p. 167, pl. 14, fig. 5, and pl. 15, fig. 3; Wulff et al., 1968, p. 58; Earle, 1969, p. 205, fig. 104; Rhodes and Connell, 1973, p. 212, figs. 2–10; Lüning and Dring, 1973, p. 334, figs. 1–6.

GENUS *PUNCTARIA*

Plant a flat blade, tapering sharply at the base to a slender stalk and attached by a basal disk. Blade four to seven cell layers thick, the surface cells not conspicuously smaller than those beneath; surface of blade with numerous tufts of hairs. Unilocular sporangia scattered; plurilocular sporangia clustered, partially immersed, the apex projecting.

Key to the Species of the Genus *Punctaria*

Blade light brown, two to four cells thick . . . . . *P. latifolia*
Blade dark brown, four to seven cells thick . . . *P. plantaginea*

PUNCTARIA LATIFOLIA GREVILLE

(**Fig. 70,** p. 170)

Plants forming lanceolate blades with a basal disk and a short stalk, 8–30 cm or more long, 1–8 cm wide, the base abruptly tapering, the apex usually rounded. Blade two to four cells (50–65 μm) thick, soft and thin-membranous, yellowish- to olive-brown, becoming more greenish on drying. Surface cells 15–40 μm in diameter, in fairly distinct rows.

In Virginia it is found on *Zostera* leaves and on stones and shells in the lower Chesapeake Bay and along the Eastern Shore during winter and spring. Known from North Carolina to Newfoundland.

Greville, 1830; p. 52; Williams, 1948, p. 689; Taylor, 1957, p. 166, pl. 15, fig. 5; Mathieson, Dawes, and Humm, 1969, p. 127.

PUNCTARIA PLANTAGINEA (ROTH) GREVILLE

(**Fig. 71,** p. 171)

Plants forming lanceolate blades to a height of about 40 cm, sometimes more, 1.0–2.5 cm wide, membranous-leathery, the base tapering to a short stalk, the apex tapering or rounded to truncate, or sometimes worn and split. Blade four to seven cells thick (110–225 μm). Cell walls fairly thick, the surface cells 15–40 μm in diameter in surface view. Plurilocular sporangia produced on the surface, oblong to ovoid, 20–35 μm in diameter, 30–50 μm long; unilocular sporangia spherical, 32–50 μm in diameter.

Known in Virginia from the breakwater at Cape Charles and from other stations in the lower Chesapeake Bay and along the Eastern Shore during winter and spring only. Atlantic coast distribution: North Carolina to Labrador and Baffin Island.

Greville, 1830, p. 53; Rosenvinge and Lund, 1947, p. 11, fig. 2; Taylor 1957, p. 166, pl. 15, fig. 4, and pl. 16, fig. 4; Zaneveld and Barnes, 1965; Zaneveld and Willis, 1976, p. 39.

GENUS *ASPEROCOCCUS*

Plants an elongate hollow tube with a short stipe and a basal attachment disk, the plant usually in groups. Innermost cells large, of a few layers, decreasing in size outward, the small cells of the surface layer with many chromatophores and bearing tufts of short, colorless hairs. Unilocular and plurilocular sporangia on the same or on different plants, produced in sori.

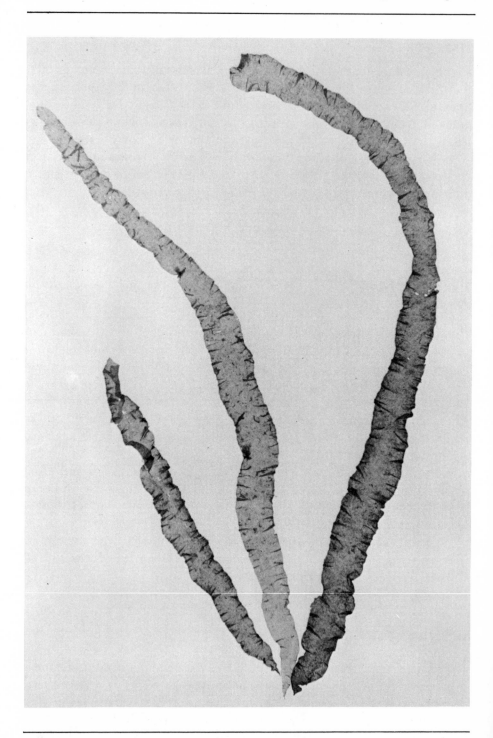

Fig. 70.   *Punctaria latifolia,* photograph of herbarium specimens (×½)

Fig. 71. *Punctaria plantaginea*, photograph of herbarium specimens (×½)

ASPEROCOCCUS FISTULOSUS (HUDSON) HOOKER

(Fig. 72)

Plants 10–45 cm tall, 2–5 mm in diameter, dark brown, not slippery. Surface cells elongate, 16–26 μm wide as seen in surface view, 16–36 μm long, 9–22 μm thick as seen in cross section. Unilocular sporangia in longitudinally elongate sori with paraphyses consisting of three or more cells; the sporangia on one-celled stalks or sessile, ovoid to spherical, 38–55 μm in diameter, 55–70 μm long.

   On rocks of the breakwater and on shells at Cape Charles, Virginia, and probably widely distributed in the lower Chesapeake Bay and along the Eastern Shore in winter and spring. Known from North Carolina to Baffin Island.

   Hudson, 1778, p. 569 (as *Ulva fistulosa*); Hooker, 1833, p. 277; Earle, 1969, p. 199 (the latter record of considerable doubt); the following as *A. echinatus* (Mertens) Greville: Greville, 1830, p. 50, pl. 9; Williams, 1948, p. 689; Taylor, 1957, p. 169, pl. 15, fig. 7, and pl. 16, figs. 5–6.

GENUS *SCYTOSIPHON*

Plants forming an unbranched hollow tube, firmly membranous, the inner cells large and cylindrical, the outer rounded or angular and with many chromatophores. Growth intercalary near the base. Plurilocular sporangia in sori or forming an expansive layer, interspersed with numerous paraphyses.

SCYTOSIPHON LOMENTARIA (LYNGBYE) J. AGARDH

(Fig. 73, p. 174)

Plants in groups, short-stalked, with a disklike holdfast, tapered-cylindrical, 15–60 cm tall or more, 3–10 mm in diameter, frequently with constrictions when mature. Plurilocular sporangia in patches, the sporangia 3.5–7.5 μm in diameter, 45–65 μm long, uniseriate or sometimes biseriate or forked. Paraphyses numerous among the sporangia, 27–32 μm long, 8–13 μm in diameter, not taller than the sporangia.

   On shells and stones at the mouth of Willoughby Bay, Hampton Roads; at Chincoteague Inlet, and generally widely distributed in the lower Chesapeake Bay and along the Eastern Shore of Virginia during

Fig. 72. *Asperococcus fistulosus,* photograph of an herbarium specimen of a group of plants ($\times\frac{1}{2}$)

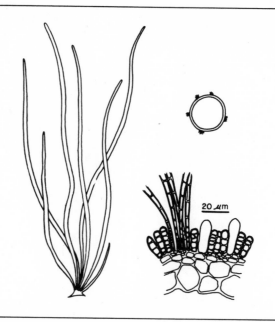

Fig. 73. *Scytosiphon lomentaria:* at left, habit sketch of a group of plants (×⅓); above right, a cross section showing groups of sporangia and paraphyses; below right, enlarged view of a portion of a cross section through a sorus showing gametangia, paraphyses, and hairs.

winter and spring. Known from South Carolina to northern Labrador.

Variety *complanatus* Rosenvinge is characterized by plants that tend to be soft, slender, and without constrictions, though often flattened. The plants are 1.5–2.0 mm in diameter, 30–60 cm tall, and without paraphyses. This variety seems to be the more common representative of the species in the southern part of the range.

Rhodes and Connell (1973) have shown that *S. lomentaria* passes the summer in Virginia waters in the form of flat crusts that resemble the brown algal genus *Ralfsia*. In the winter, these crusts give rise to typical *Scytosiphon* plants. No evidence of sexual reproduction has been observed in Virginia.

Hollenberg and Abbott (1966) established the family Scytosiphonaceae for members of the genus *Scytosiphon*, removing them from the family Punctariaceae (or Encoeliaceae) because of the fact that they exhibit a definite basal meristematic area. Members of the family Punctariaceae would thus be restricted to genera with intercalary growth throughout the plant.

J. Agardh, 1848–76 (1848), p. 126; Blomquist and Humm, 1946, p. 6; Rosenvinge and Lund, 1947, p. 27, fig. 9; Taylor 1957, p. 168, pl. 15, fig. 2, and pl. 16, fig. 3; Rhodes and Connell, 1973, p. 212, figs. 11–14; Lüning and Dring, 1973, p. 336, figs. 13–18.

FAMILY DICTYOSIPHONACEAE

Plants growing from an apical cell, slender, the branching irregular, bearing delicate hairs on the branches. Unilocular sporangia produced by transformation of surface cells, the spores producing filamentous, microscopic gametophytes that are isogamous.

GENUS *DICTYOSIPHON*

Plants with a distinct main axis, from which many branches arise in an irregular fashion. Axes in young plants with a medulla of large cells that is covered by a thin cortex of small cells, the axes becoming hollow in old plants.

Key to the Species of the Genus *Dictyosiphon*

Plants 2–8 cm tall, unbranched or with few branches . . . . . .
. . . . *D. eckmani*

Plants 20–70 cm tall, much branched . . . . . *D. foeniculaceus*

DICTYOSIPHON ECKMANI ARESCHOUG

Plants small and delicate, 2–8 cm tall but sometimes only a few millimeters in height, unbranched or sparingly branched at the top, the axis less than 1 mm in diameter, widest near the center. Sporangia relatively large, 22–34 μm in diameter in surface view. An epiphyte of *Scytosiphon* and *Petalonia*.

Found in Virginia by Dr. Russell Rhodes. Previously known from the north side of Cape Cod to Nova Scotia.

Areschoug 1846–50; Kylin, 1947*a*, p. 79, pl. 14, fig. 45; Edelstein and McLachlan, 1966, p. 1050, fig. 36.

DICTYOSIPHON FOENICULACEUS (HUDSON) GREVILLE

(Fig. 74)

Plants filiform, firm, much-branched, 20–70 cm tall or more, the axes no more than 0.5 mm in diameter; delicate hairs abundant on the

Fig. 74.   *Dictyosiphon foeniculaceous,* photograph of an herbarium specimen showing many plants growing upon *Scytosiphon lomentaria* ($\times\frac{1}{2}$)

younger branches. Unilocular sporangia immersed in the surface cell layer, broadly oval, 30–55 μm in diameter, the only reproductive structures known.

Reported for Virginia by Rhodes (1970a) in Burtons Bay near Wachapreague on the Eastern Shore during winter and spring. Previously known from New Jersey to Newfoundland, Baffin Bay, and the Arctic.

Hudson, 1762, p. 479 (as *Conferva foeniculacea*); Greville, 1830, p. 56, pl. 8; Rosenvinge and Lund, 1941, p. 63, fig. 22; Taylor, 1957, p. 172, pl. 12, fig. 4, and pl. 14, fig. 2; Zaneveld and Willis, 1976, p. 40.

ORDER FUCALES

Large plants of parenchymatous structure in which the gametophyte phase is reduced to the gametangium and gametes; oogamous, the oogonia borne within special cavities (conceptacles) either scattered over the surface of the plant or restricted to special branches or branch tips (receptacles).

There are two families represented in the Virginia flora. The Fucaceae, represented by *Fucus* and *Ascophyllum,* is a cool- or cold-water group, members of which are mostly restricted to the intertidal zone. The Sargassaceae, represented by *Sargassum,* is a family of warm or tropical waters, members of which are mostly found below low tide, some at considerable depths. One member of the Fucaceae is found as far south as North Carolina (*Fucus vesiculosus*); one member of the Sargassaceae is found as far north as the south side of Cape Cod (*Sargassum filipendula*).

FAMILY FUCACEAE

Plants dichotomously or pinnately branched, subterete to costate, some with piliferous cryptostomata, and often with air bladders. Receptacles produced seasonally in most genera, terminating main branches, or on short lateral branchlets, heterosporus, the oogonia producing one to eight functional eggs. This family represents the cold-water members of the order Fucales.

GENUS *ASCOPHYLLUM*

Plants pinnately branched, without a midrib or cryptostomata, tough-cartilaginous. Receptacles produced in pinnately arranged short pedicels that are deciduous. Oogonia producing four eggs.

ASCOPHYLLUM NODOSUM (L.) LE JOLIS

(Fig. 75)

Plants olive, yellowish-brown, or brown, usually 30–50 cm tall, attached by a disk-shaped holdfast, the main axis and branches flattened, without a midrib, and bearing large, single air bladders. Branching irregular to pinnate, the ultimate branches short, simple or forked, pinnately disposed, 1–2 cm long, these developing into or replaced by the deciduous, yellowish receptacles borne on stalks singly or in clusters of a few.

Known attached from Delaware to the Arctic, usually intertidal. Apparently *Ascophyllum* does not occur in Virginia except in the form of loose, drifting plants. These, however, are common along the Eastern Shore and occasional in the lower Chesapeake Bay. Since this species has air bladders, and floats when dislodged from its place of attachment, it often drifts hundreds, even thousands, of miles. Plants are carried southward on the inshore, open-sea current from Cape Cod to Cape Hatteras. They are often encountered mixed with the pelagic *Sargassum* species in the Sargasso Sea along with species of *Fucus*.

It seems strange that during winter in Virginia waters drifting plants do not produce viable embryos that would become attached, especially in the vicinity of Chincoteague Island. The unattached and drifting condition, however, often tends to inhibit reproduction in the algae, and this may apply to *Ascophyllum,* even when the water temperature and other conditions are favorable to gamete production.

Le Jolis, 1856, p. 83; Farlow, 1881, p. 99; Taylor, 1957, p. 195; Zaneveld and Willis, 1976, p. 42.

GENUS *FUCUS*

Plants intertidal, attached by a disk-shaped holdfast, erect, usually dichotomously branched, the branches strap-shaped, with a midrib, the lower branches often consisting of the midrib alone as a result of decaying of the alate portion. Air bladders usually present, often in pairs. Cryptostomata usually present on sterile parts of the blades; a group of colorless hairs protruding from the cryptostomata. Receptacles terminal on main or lateral branches, seasonal, much inflated. Oogonia producing eight eggs.

Key to the Species of the Genus *Fucus*

Receptacles flattened, four to ten times as long as broad; midrib scarcely visible in upper branches; air bladders absent　.　.　.　*F. edentatus*

Fig. 75. *Ascophyllum nodosum,* photograph of an herbarium specimen showing part of a plant

Receptacles inflated, one and a half to two times as long as broad; midrib distinct throughout; air bladders present . . . *F. vesiculosus*

FUCUS EDENTATUS DE LA PYLAIE

(Fig. 76)

Plants 20–40 cm tall, attached by a broad, conical disk, the branching dichotomous with narrow angles. Midrib distinct below, indistinct above, but toward the base the midrib only may be present as a result of loss of the alate portion. Air bladders absent; cryptostomata small and inconspicuous. Receptacles much elongated, broadest at the base, tapering upwards into points, often forked, four to ten times as long as they are broad and not sharply distinct from the sterile portion of the branch.

Willis (1973) found this species attached to shells at Smith Island, Virginia. It had previously been known in the state only as drifting plants and had been reported attached from New Jersey to Newfoundland. Drifting plants have been collected many times along the Eastern Shore, at Lynnhaven Inlet, and Virginia Beach, especially during the winter months when prevailing winds are north or northeast.

De la Pylaie, 1829, p. 84, pl. 23; Farlow, 1881, p. 102 (as *F. furcatus* C. Agardh); Taylor, 1957, p. 191, pl. 23, fig. 3; Zanevelt and Willis, 1976, p. 42.

FUCUS VESICULOSUS LINNAEUS

(Fig. 77, p. 182)

Plants intertidal, 20–50 cm tall, attached by an irregular, lobed holdfast, the branching dichotomous or somewhat irregular, the angles not as narrow as in *F. edentatus*. Midrib prominent throughout, the lower parts of branches often consisting of the midrib only. Air bladders usually in pairs and much swollen. Cryptostomata scattered but distinct. Receptacles terminal on the branches, single, in pairs, or forked, usually no more than twice as long as they are wide, inflated, sharply distinct from the sterile portion of the branch.

Abundant on rocks and in salt marshes along the Eastern Shore of Virginia, and occasional in attached condition in the Hampton Roads area and elsewhere in the lower Chesapeake Bay. Known to grow attached from North Carolina to the Arctic, the southernmost species of the family Fucaceae.

Linnaeus, 1753, p. 1158; Farlow, 1881, p. 100, pl. 9; Hoyt, 1917–18, p. 450, pl. 89; Taylor, 1957, p. 192, pl. 25, figs. 1–3.

Fig. 76. *Fucus edentatus,* photograph of an herbarium specimen (×½)

Fig. 77. *Fucus vesiculosus,* photograph of an herbarium specimen

FAMILY SARGASSACEAE

Plants with monopodial branching, though somewhat obscure in the pelagic species, the branches costate or differentiated into leaflike and stemlike organs, bilateral or radial in organization. Cryptostomata and air bladders usually present. Reproduction oogamous; gametes liberated into the water or the eggs held by a coating and strand of algin for a few days, with liberation of the microgametes only. Where eggs are held on the outside of the conceptacle after extrusion, they are released as multicellular embryos. Gametogenesis and release may be periodic in most species.

GENUS *SARGASSUM*

Plants attached by well-developed holdfasts or free and strictly pelagic; bushy, usually bearing distinct, terete branches with narrow to broad leaflike appendages and air bladders. Receptacles, conceptacles, and cryptostomata produced by the attached species, the receptacles usually axillary and branched. A single functional egg is produced in meiotic division of the megagametangium, with seven nuclei degenerating.

Key to the Species of the Genus *Sargassum*

1. Never attached; without receptacles and cryptostomata, strictly pelagic 2

1. Always attached, at least at first; cryptostomata present, receptacles and conceptacles produced . . . . *S. filipendula*

2. Leaves narrow, linear; the marginal teeth long-tipped; vesicles often apiculate . . . . *S. natans*

2. Leaves linear to lanceolate; marginal teeth not spine-tipped; vesicles rarely apiculate . . . . *S. fluitans*

SARGASSUM FILIPENDULA C. AGARDH

(**Fig. 78**)

Plants 30–60 cm tall, attached by a large holdfast, the branches bearing alternate, petiolate leaves, spherical axillary air bladders, and, when fertile, cervicorn or racemosely branching receptacles arising in leaf axils in the upper parts of the plant. Leaves linear-lanceolate, sometimes forked, serrate, with a distinct midrib and numerous, scattered cryptostomata; leaves 3–8 cm long, 5–8 mm wide, largest at the base of the plant. Air bladders 3–5 mm in diameter, on stalks about 5 mm long, sometimes apiculate. Receptacles bisexual, producing gametes at regular intervals during the growing season, the eggs ordinarily remaining at-

Fig. 78. *Sargassum filipendula,* photograph of an herbarium specimen

tached to the outside of the conceptacle by a coating and strand of algin for one to several days after extrusion. Fertilization occurs at the time of extrusion, usually at daybreak, and development of an embryo begins immediately, the embryo usually dropping off after 24 to 48 hours in a many-celled condition and with polarity established. The embryo is briefly planktonic.

S. *filipendula* occurred on the rocks of the breakwater at Cape Charles in the early 1940s, but specimens were not made. Apparently it has not been seen again attached in Virginia waters and may not be a member of the flora at the present time. Drifting plants are occasionally seen along the Eastern Shore, and they probably come into Virginia waters from the north, as the species is known attached from the Caribbean Sea to North Carolina, and north of Virginia to Buzzards Bay on the south side of Cape Cod.

Variety *montagnei* (Bailey) Collins and Hervey is characterized by linear leaves, frequently apiculate or mucronate bladders, and by long, slender, less-branched receptacles. It is typically a plant of deeper water and open-sea conditions and has been recorded from the Caribbean Sea to Cape Cod also, but it is not often collected attached. It may occur attached on wrecks or other solid substrata in ten to thirty meters of water off the open sea coast of Virginia. Drifting plants are occasional on the outer beaches of Virginia.

C. Agardh, 1824, p. 300; Farlow, 1881, p. 103; Collins and Hervey, 1917, p. 83; Wood and Palmatier, 1954, p. 140; Taylor, 1957, p. 197, pl. 27, figs. 4–6; Taylor, 1960, p. 270, pl. 37, fig. 3, and pl. 40, fig. 2; Humm and Caylor, 1957, p. 249, pl. 8, fig. 7; Earle, 1969, p. 217, fig. 117; Zaneveld and Willis, 1976, p. 42.

SARGASSUM NATANS (L.) J. MEYEN

(Fig. 79)

Plants strictly pelagic, never attached, always sterile, without receptacles or cryptostomata, native of the Sargasso Sea and often abundant in the Caribbean Sea and the Gulf of Mexico. Plants branched in all directions, the main axis usually obscure, stems smooth. Leaves narrow-linear, 2.5–5.0 cm long, 2.0–3.5 mm wide, serrate, the teeth slender and extended. Air bladders 3–5 mm in diameter on 3–5 mm stalks, apiculate or spine-tipped or sometimes with a reduced leaf at the apex.

Often washed ashore along the outer coast of Virginia, especially in the summer months. This plant drifts ashore along the Atlantic coast of North America from Florida to Nova Scotia and Newfoundland.

Linnaeus, 1753, p. 1160 (as *Fucus natans*); Meyen, 1838, p. 185;

Fig. 79. *Sargassum natans,* photograph of an herbarium specimen

Farlow, 1881, p. 103 (as *S. bacciferum*) ; Humm and Caylor, 1957, p. 249, pl. 9, fig. 1; Taylor, 1960, p. 281, pl. 37, fig. 2, and pl. 40, figs. 3, 8; Earle, 1969, p. 225, fig. 121.

SARGASSUM FLUITANS BØRGESEN

(Fig. 80)

Plants strictly pelagic, never attached, always sterile, without receptacles and usually without cryptostomata. A few cryptostomata may occasionally be seen. Native of the Sargasso Sea and often abundant in the Caribbean Sea and the Gulf of Mexico, mixed with *S. natans*. Plants widely branched, the main axis usually obscure, the stems smooth or occasionally sparingly spinose. Leaves lanceolate, 2–6 cm long, 3–8 mm wide, serrate, the teeth flattened, not extended. Air bladders oval to subspherical, 4–5 mm in diameter, on a short stalk, the tips not apiculate or rarely with an appendage.

Often washed ashore along the outer coast of Virginia, especially in the summer months; usually mixed with *S. natans*. Like the latter, this species also drifts ashore along the Atlantic coast of North America from Florida to Nova Scotia and Newfoundland.

Børgesen, 1914, p. 66; Humm and Caylor, 1957, p. 249, pl. 9, fig. 1; Taylor, 1960, p. 281, pl. 39, fig. 2, and pl. 40, fig. 7; Earle, 1969, p. 221, fig. 120; Zaneveld and Willis, 1976, p. 42.

Fig. 80.   *Sargassum fluitans,* photograph of an herbarium specimen

CHLOROPHYTA

Green algae, Chlorophyta, represent, in part, ancestral types from which the land plants are believed to have been derived, including the mosses, ferns, and seed plants. Green algae have essentially the same photosynthetic and other pigments as the land plants: chlorophylls *a* and *b*, beta-carotene, and several other carotenes and xanthophylls. Starch is the principal form of food storage. Land plants and green algae both have cellulose and certain other polysaccharides in their cell walls. The differences between the two groups in this respect are derived from adaptations to the air-soil media of the one and the aqueous medium of the other. Land plants produce wood by lignification of cellulose, a biochemical pathway that provides stems with sufficient rigidity to stay erect in air. Plants of aquatic origin, having different environmental problems, produce instead a variety of gelatinous polysaccharides in the cell walls.

Sexual reproduction in green algae involves gametes or zoospores produced in unicellular gametangia or sporangia or in a coenocytic segment of the plant. In those species having separate gametophyte and sporophyte stages, the gametophytes produce gametangia and gametes, the sporophytes produce sporangia in which meiosis occurs. The zygotes germinate into sporophytes; the zoospores germinate into gametophytes. Gametes and spores of this type are an equal and opposite part of the sexual process, the former leading to syngamy, the latter to meiosis. In terms of gamete morphology, green algae may be isogamous, anisogamous, or oogamous. Motile cells usually have two or four flagella at the anterior end.

Asexual reproduction is commonly carried on by the production of nonmotile spores (aplanospores or akinetes) or motile spores (zoospores) having the same chromosome complement as the parent plant. Fragmentation of the plant may also occur.

Gametophyte and sporophyte stages may be morphologically identical (isomorphic), or they may be slightly to extremely unlike morphologically (heteromorphic). In other green algae, all plants are gametophytes and the sporophyte is represented only by the zygote; or, the plants are sporophytes and the gametophyte is represented only by the gametes.

The characteristics of the life cycle in green algae are of little phylogenetic value in the general classification; that used here is based upon Smith, 1955.

Green algae are widely distributed in both fresh and salt water. There are about 6,500 species in 425 genera (Smith, 1955), of which only about 10% are marine. This statement, if made without explanation, is misleading to a student who has not had an opportunity to see the green algae of marine water, especially in the tropics. Far from being insignificant in the sea, they make up an important segment of marine benthic vegetation. Nearly all of the large, highly developed Chlorophyceae are marine, those belonging to the orders Siphonocladiales, Dasycladales, and Siphonales. Members of the Siphonales, especially the genera *Caulerpa, Halimeda, Penicillus, Avrainvillea,* and *Udotea,* are the only benthic algae in the sea that have developed the ability to colonize unconsolidated bottom sediments. These plants grow in seagrass beds or in sand or muddy sand where other algae cannot attach.

The genera of green algae listed above, as well as certain groups of brown and red algae, are perhaps as highly evolved, as morphologically complex in relation to their environment, as land plants. A plant of this complexity is something more than a "thallus."

CLASS CHLOROPHYCEAE

Plants unicellular, multicellular, of coenocytic segments, or wholly coenocytic. Because of their aquatic environment, they do not have ensheathing structures around the gametangia and they do not develop flowers, fruits, or seeds.

All true green algae, except the stoneworts, are included in this class. Smith (1955) placed the stoneworts in the class Charophyceae. Silva (1962) placed them in a division, the Charophyta. They are not treated in this work, as they are primarily freshwater plants.

Key to the Genera or Species of the Green Algae

The following key leads to the genus or, in the case of those genera represented in this work by a single species, to the species. Keys to those genera with two or more species will be found following the genus description of each.

1. Individual plants microscopic, through the plant mass may be readily visible 2

1. Individual plants macroscopic 10

2. All or most of the plant boring into shells or limestone . . . 3

2. Plants not penetrating limestone 4

3. Filaments with cross walls . . *Gomontia polyrhiza* (p. 201)

3. Filaments without cross walls, irregular . . . .
. . . . *Ostreobium quekettii* (p. 126)

4. Filaments uniseriate, unbranched 5

4. Filaments branched, or plants of coccoid cells forming a one-celled layer 6

5. Cells short, usually less than one diameter long; one nucleus . .
. . . . *Ulothrix* (p. 193)

5. Cells usually over one diameter long; multinucleate . . . .
. . . . *Rhizoclonium* (p. 220)

6. Plants composed of a layer of coccoid cells, slightly filamentous at margin of layer . . . . . *Protoderma marinum* (p. 200)

6. Plants of uniseriate or coenocytic branched filaments . . . 7

7. Plants of creeping filaments that give rise to erect filaments 0.5–2.0 mm tall . . . . *Pilinia rimosa* (p. 196)

7. Plants of prostrate filaments only 8

8. Coenocytic, producing a cushion or turf in bottom sediments . .
. . . . *Vaucheria thuretii* (p. 127)

8. Cellular and epiphytic 9

9. Some cells bearing erect, colorless, slightly spiral hairs . . .
. . . . *Phaeophila dendroides* (p. 197)

9. Without hairs; plants within the surface polysaccharide of the host plant or within the bryozoan *Alcyonidium* . . . . . .
. . . . *Entocladia* (p. 198)

10. Plants of uniseriate or biseriate filaments, with cross
walls 11

10. Plants forming a hollow tube, a flat sheet, or
coenocytic 15

11. Filaments biseriate, unbranched . . . .
. . . . *Percursaria percursa* (p. 202)

11. Filaments uniseriate 12

12. Densely branched, the branches similar to the main axes . . .
. . . . *Cladophora* (p. 224)

12. Unbranched (or with occasional rhizoidal branches
only) 13

13. Filaments usually over 100 μm in diameter; cells barrel-shaped or
with constrictions at nodes . . . . *Chaetomorpha* (p. 218)

13. Filaments less than 100 μm in diameter 14

14. Plants attached in exposed places; without rhizoidal branchlets .
. . . . *Urospora mirabilis* (p. 218)

14. Plants loose, entangled; in salt marshes and protected places;
often with rhizoidal branchlets . . . *Rhizoclonium* (p. 220)

15. Plants composed of a hollow tube of cells, the tube sometimes collapsed and evident only at the base . . . .
. . . . *Enteromorpha* (p. 202)

15. Plants a cellular flat sheet or coenocytic 16

16. Plants a flat, wide sheet 17

16. Plants coenocytic 18

17. Sheet one cell thick . . . . *Monostroma* (p. 212)

17. Sheet two cells thick . . . . *Ulva* (p. 215)

18. Plants bushy-branched, attached to a solid substrate; branchlets
usually distichous . . . . *Bryopsis* (p. 230)

18. Plants growing in bottom sediments, a turf or cushion . . . .
. . . . *Vaucheria thuretii* (p. 127)

ORDER ULOTRICHALES

Plants of unbranched or branched filaments or foliaceous; cells uninucleate, with usually a single band-shaped, parietal chromatophore and
one to several pyrenoids. Sexual reproduction by zoospores (if meiosis
occurs in the sporangium) and isogamous, anisogamous, or oogamous
gametes. Asexual reproduction by neutral zoospores, aplanospores, or
akinetes.

FAMILY ULOTRICHACEAE

Plants uniseriate, typically unbranched, unattached or with a modified
basal holdfast cell; chromatophore platelike or bandlike, with or without pyrenoids. Gametes isogamous or anisogamous.

GENUS *ULOTHRIX*

Plants uniseriate, unbranched, originally attached by a basal holdfast
cell, often becoming free and entangled. Cells with one bandlike chromatophore parietal in position and more or less encircling the cell, with
one to three pyrenoids. Zoospores quadriflagellate. Gametes biflagellate.

Key to the Species of the Genus *Ulothrix*

1. In the outer layer of sponges . . . . *U. endospongialis*
1. Plants free-living 2

2. Cells isodiametric or shorter, the chromatophore filling the cell
    . . . . *U. flacca*
2. Cells mostly longer than wide, the chromatophore only partly filling
    the cell . . . . *U. subflaccida*

ULOTHRIX ENDOSPONGIALIS N.SP.
(Fig. 81)

*Filamenta diametro 4–6.5 μm non ramosa, ad dissepimenta non con-
stricta, per textum superficialem superiorem spongiae Microcioniae pro-
liferae sinuantia, spongiam nigrescenti-viridem colorantia, infra aestum
medio-inferiorem in pila atque conchas ostryarum. Cellulae 2–4 diame-
tros longae, fortuitae tumidae diametro ad 7 μm, chromatophora sin-
gulo, laminam parietalem sistente, per longitudinem cellulae extendente,*

Fig. 81. *Ulothrix endospongialis,* a segment of a filament

anulum non integrum formati, pyrenoidea unica. Cellula terminalis non attenuans. Filamenta juvenilia discos affigentes basales vel passim laterales in texto spongiae portantia.

Ramifying throughout the upper parts of *Microciona prolifera,* an orange-colored sponge that grows on pilings and oyster shells about six inches below mean low tide, and coloring the sponge blackish-green. Filaments 4.0–6.5 μm in diameter, unbranched; the cells 2–4 diameters long, uninucleate; the chromatophore a parietal plate occupying essentially the entire length of the cell but not forming a complete ring, with one pyrenoid, sometimes two. Cells not constricted at the nodes, terminal cells not tapered; occasional cells swollen to 7 μm in diameter. Young filaments showing basal and sometimes lateral attachment, but evidently always within the sponge tissue. Reproduction not observed.

Culture studies of this plant are needed to determine its relationship to the sponge and to other species of the genus *Ulothrix.*

York River at Yorktown and Gloucester Point, the type from old pilings of the former ferry dock at Gloucester Point; probably widely distributed in the Chesapeake Bay, as determined by salinity. Year around.

The type specimen has been deposited in the herbarium of the Virginia Institute of Marine Science.

ULOTHRIX FLACCA (DILLWYN) THURET
(Fig. 82)

Filaments 10–25 μm in diameter, the cells shorter than wide (mostly 4–15 μm long); chromatophores occupying all or most of the face of

Fig. 82.   *Ulothrix flacca*, filaments showing reproduction

the cell, with one to three pyrenoids. Reproductive cells much swollen. Plants often loose and entangled or wrapping around invertebrate animals on pilings.

Reported for Virginia by Wulff and Webb (1969) on old pilings at Gloucester Point, from just below mean low water to about 45 cm above, and from a jetty at Chincoteague Inlet by Dr. F. D. Ott; both collections during winter or spring. Known from Texas to Newfoundland and Baffin Island.

Thuret in LeJolis, 1863, p. 56; Newton, 1931, p. 56, fig. 40; Blomquist and Humm, 1946, p. 4, fig. 10; Taylor, 1957, p. 45, pl. 1, fig. 9; Edwards, 1970, p. 16, fig. 19; Chapman, 1971, p. 81; Zaneveld and Willis, 1974, p. 66.

ULOTHRIX SUBFLACCIDA WILLE

Filaments unbranched, 5–25 μm in diameter, attached by a rounded basal cell; the cells from about one to two diameters long; the band-shaped chromatophore with one pyrenoid and not occupying the entire cell.

Known from Cuba and the Florida Gulf coast. Found on oyster shells at Gloucester Point, Virginia, by Sybil Ramsey.

Wille, 1901, p. 27, pl. 3, figs. 90–100; Collins, 1909, p. 186; Kylin, 1949, p. 13; Humm and Jackson, 1955, p. 241; Humm and Taylor, 1961, p. 333, fig. 3a.

FAMILY CHAETOPHORACEAE

Plants filamentous, branched, composed of a basal layer of creeping filaments from which erect filaments are produced or of creeping filaments only; uniseriate. In some genera the plants form flat disks because of the close association of the horizontal filaments. Colorless hairs are often present, either terminal or lateral. Cells uninucleate, with a single platelike chromatophore.

GENUS *PILINIA*

Plants of branched, creeping filaments that give rise to erect, unbranched or sparingly branched filaments that are densely aggregated, many bearing a multicellular hair at the tip. Chloroplasts without pyrenoids. Reproduction by biflagellate zoospores produced in terminal or lateral, oval to clavate sporangia.

PILINIA RIMOSA KÜTZING
(Fig. 83)

Plants producing a firmly attached, green coating on the substratum or host 0.5–2.0 mm tall. Basal, creeping filaments somewhat irregular and

Fig. 83. *Pilinia rimosa,* a portion of a plant showing the basal layer, a sporangium, and the sparingly branched erect filaments with hair tips

often constricted at the nodes; erect filaments branched or unbranched, the cells 5–15 μm in diameter, 1–2 diameters long, cylindrical or sometimes barrel-shaped. Sporangia 15–20 μm in diameter.

In Virginia, this species was collected by Dr. Barry Wulff on the old ferry-dock pilings at Gloucester Point, October 10, 1966. It has been recorded in the western Atlantic Ocean only a few times. Collins (1909) was the first; he found it in Maine and distributed specimens as Phycotheca Boreali Americana No. 971. Humm and Taylor (1961) recorded it from the Florida Gulf coast; Zaneveld and Willis (1974) found it in Delaware.

Kützing, 1843, p. 273; Collins, 1909, p. 291; Hamel, 1931, p. 37; Taylor, 1957, p. 48; Zaneveld, 1972, p. 129.

GENUS *PHAEOPHILA*

Plants of prostrate, branching filaments, epiphytic or endophytic; occasional cells with long, colorless hairs that are nonseptate and not separated from the supporting cell by a wall. Cells with lobed, parietal chromatophores and two or three pyrenoids. Reproduction by quadriflagellate zoospores, the zoosporangia not much enlarged, often terminal on branches.

PHAEOPHILA DENDROIDES (CROUAN) BATTERS

(Fig. 84)

Filaments creeping on the surface or within the polysaccharide layer of various algae, seagrasses, animals, shells, and wood; irregularly branched. Cells cylindrical to irregular in shape, sometimes swollen, mostly 10–25 μm in diameter, 1–2 diameters long. Occasional cells bearing long, colorless erect hairs that are slightly spiral and hence appear to be zigzag. Zoosporangia intercalary or terminating short branches, usually somewhat swollen, 15–40 μm in diameter, 30–85 μm long.

Common on larger algae of the higher salinity area of the lower Chesapeake Bay and along the Eastern Shore. Known from the Caribbean Sea, Gulf of Mexico, and Florida to Nova Scotia.

Crouan and Crouan, 1867, p. 128, pl. 8, fig. 59 (as *Ochlochaete dendroides*); Batters, 1902, p. 13; Humm and Caylor, 1957, p. 242, pl. 5, fig. 1 (as *P. floridearum* Hauck); Edelstein and McLachlan, 1967*b*, p. 214, figs. 5–6.

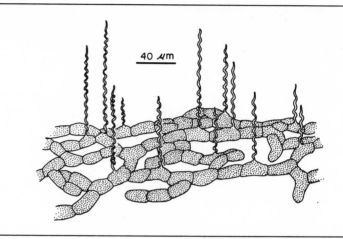

Fig. 84. *Phaeophila dendroides,* a portion of a plant bearing erect, colorless, spiral hairs

GENUS *ENTOCLADIA*

Plants forming small to extensive green patches on or in the host; the filaments spreading, branched; cells without hairs or setae, with one platelike chromatophore having one to three pyrenoids; uninucleate. Asexual reproduction by quadriflagellate zoospores; sexual reproduction by biflagellate isogametes.

Key to the Species of the Genus *Entocladia*

Filaments 3–8 µm in diameter, often irregular and contorted . . . . *E. viridis*

Filaments 5–10 µm diameter, mostly regular and cylindrical . . . . *E. wittrockii*

ENTOCLADIA VIRIDIS REINKE
(Fig. 85)

Plants creeping within the polysaccharide coating of the host or within the gelatinous portion of the bryozoan *Alcyonidium verrilli* and coloring the host blackish-green. Cells mostly 3–6 µm in diameter, 1–6 diameters long, cylindrical to irregularly swollen.

Common in larger algae and *Alcyonidium* at Hampton Roads, Lynhaven Inlet, and other areas of the Chesapeake Bay and the Eastern Shore of Virginia. Known from the tropics to Nova Scotia.

Reinke, 1879, p. 476, pl. 6, figs. 6–9; Taylor, 1957, p. 54, pl. 2,

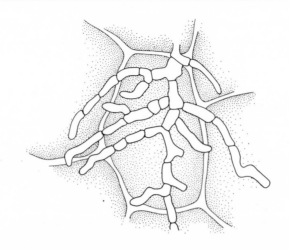

Fig. 85. *Entocladia viridis,* filaments creeping within the polysaccharide layer of the host plant

figs. 1–2; Edelstein and McLachlan, 1966, p. 1046; p. 39; Cardinal, 1968, p. 39; Zaneveld and Willis, 1974, p. 66.

ENTOCLADIA WITTROCKII WILLE

Plants creeping within the polysaccharide surface layer of larger algae, irregularly branched, the branches sometimes united laterally, the tips tapering. Cells cylindrical, 5–10 μm in diameter (averaging 9 μm), 1–1.5 diameters long, with one pyrenoid. Apparently more common on brown than on red or green algae.

On larger algae in bays along the Eastern Shore of Virginia near Wachapreague, summer of 1962. Known from the Gulf of Mexico and Florida to Newfoundland.

Wille, 1880, p. 3, pl. 1, figs. 1–7; Collins, 1909, p. 279, fig. 100 (as *Endoderma wittrockii*); Taylor, 1957, p. 54; Humm and Taylor, 1961, p. 338; Humm and Hildebrand, 1962, p. 237; Edelstein and McLachlan, 1967*b*, p. 213; Cardinal, 1968, p. 39; Mathieson, Dawes, and Humm, 1969, p. 117.

GENUS *PROTODERMA*

Plants forming patches or sometimes a band in the intertidal zone one cell layer thick, the cells in parenchymatous arrangement except at the margins, where there is a suggestion of filamentous arrangement. Cells without hairs or setae, with a simple, platelike chromatophore having one pyrenoid. Asexual reproduction by aplanospores or biflagellate zoospores, four or eight in a cell.

PROTODERMA MARINUM REINKE
(**Fig. 86**)

Plants forming an extensive layer of cells on rock and wood in pseudo-parenchymatous arrangement, but somewhat filamentous and radial at the margins, cells 6–12 μm in diameter.

Fig. 86. *Protoderma marinum,* a group of cells in which the filamentous nature is obscured by age

On oyster shells, stones, and a seawall in the intertidal zone of the York River near Gloucester Point, and on the carapace of a living blue crab. Probably continuous from Florida to Nova Scotia and Newfoundland.

Reinke, 1889–92 (1889), p. 81; Taylor, 1957, p. 58; Humm and Taylor, 1961, p. 341, fig. 4; Edelstein and McLachlan, 1966, p. 1046; Cardinal, 1968, p. 45.

FAMILY GOMONTIACEAE

Plants of branched, uniseriate filaments that penetrate shells. Cells with parietal chromatophores and sometimes multinucleate. Enlarged sporangia formed near the surface and becoming separated from the vegetative filaments, developing rhizoidal extensions toward the surface.

GENUS *GOMONTIA*

Plants penetrating shells and wood; of creeping, branched filaments, the cells cylindrical to irregular. Chromatophore lobed or reticulate. Cells with one to several nuclei.

Abbott and Hollenberg (1976) place *Gomontia polyrhiza* under "Algae of uncertain relationship," as it has been reported to be a stage in the life history of certain species of *Monostroma*.

GOMONTIA POLYRHIZA (LAGERHEIM) BORNET AND FLAHAULT

(Fig. 87)

Plants producing a greenish tinge on shells, the filaments 4–8 μm in diameter, irregularly branched. Vegetative cells uninucleate. Sporangia

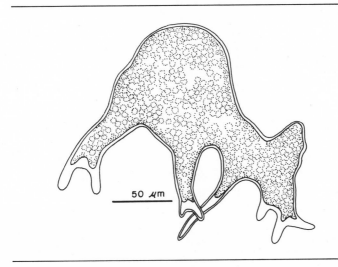

50 μm

Fig. 87. *Gomontia polyrhiza,* a large sporangium

massive, 30–125 μm in diameter, 150–250 μm long. Two sizes of zoospores, or aplanospores, may be produced.

Common in shells in the intertidal zone and below in the York River near Gloucester Point, Virginia; probably widely distributed along the Virginia coastlines. Known from the Caribbean Sea to the Arctic.

Bornet and Flahault, 1888, p. 163; Bornet and Flahault, 1889, p. 153, pls. 6–8; Hamel, 1931, p. 34, fig. 11; Taylor, 1957, p. 59, pl. 1, figs. 13–14, and pl. 7, fig. 4; Wilce, 1959, p. 65; Cardinal, 1968, p. 39; Humm, 1976, p. 85.

FAMILY ULVACEAE

Plants of slender filaments, hollow tubes, or sheets; the cells uninucleate with one or two lateral chromatophores. Gametophyte and sporophyte isomorphic. Vegetative cells develop into reproductive cells with little or no enlargement. Biflagellate or quadriflagellate zoospores are produced or biflagellate isogametes or anisogametes.

GENUS *PERCURSARIA*

Plants of slender filaments, rarely branched, usually of two rows of rectangular cells that are side by side and have a single, parietal chromatophore.

PERCURSARIA PERCURSA (C. AGARDH) J. AGARDH

Filamentous, mostly of two rows of cells, each row 10–15 μm wide, the cells 15–28 μm long; the plants forming twisted, entangled skeins or masses, especially in salt marshes or other brackish-water areas.

Found in Virginia on bent-over *Spartina* in a salt marsh in Chincoteague Inlet, March 10, 1972, by Dr. F. D. Ott. Known previously from New Jersey to Newfoundland.

C. Agardh, 1820–28, p. 424 (as *Conferva percursa*); Collins, 1909, p. 197 (as *Enteromorpha percursa*); Newton, 1931, p. 76, fig. 56; Taylor, 1957, p. 61, pl. 1, fig. 8.

GENUS *ENTEROMORPHA*

Plants usually in the form of a hollow tube, branched, or unbranched, the wall one cell thick. The tubes may be inflated or collapsed and flattened, the two layers sometimes adherent except at the margins. The branch tips often end in a solid cylinder of cells or, in some, in a uniseriate or pluriseriate row of cells. Holdfasts formed by downgrowth of rhizoidlike extensions arising from cells of the stipelike portion. Chloroplast a single, lateral plate, usually with one pyrenoid. Reproduction by zoospores or by isogametes or anisogametes formed from the vegetative cells.

The genus is worldwide and exceptionally eurythermal and euryhaline. Some species occur in fresh water and in isolated salt lakes.

Because of its morphological variation in response to environmental conditions, *Enteromorpha* has always been a difficult genus in which to delimit species. Kapraun (1970) studied the genera *Ulva* and *Enteromorpha* in the vicinity of Port Aransas, Texas, involving both field observations and laboratory culture. For the most part, his work has supported the traditional species concepts.

Key to the Species of the Genus *Enteromorpha*

1. Plants unbranched, or rarely proliferous 2

1. Plants branched, at least at the base 5

2. Plants forming lanceolate, flat blades, the two layers of cells adhering except at the margins. . . . . . *E. linza* (p. 204)

2. Plants tubular, though they may be flattened 3

3. Plants only 1–3 cm tall 4

3. Plants 10–30 cm tall, inflated, often bullate, cells 9–15 μm in diameter, not in longitudinal rows . . *E. intestinalis* (p. 205)

4. Cells in longitudinal rows at margins . . . . *E. marginata* (p. 206)

4. Cells not in longitudinal rows . . . . . *E. minima* (p. 206)

5. Branches ending in long uniseriate tips . . . . *E. plumosa* (p. 207)

5. Branches not ending in long uniseriate tips 6

6. Branches biseriate, in contrast to pluriseriate, filiform main axes . . . . *E. torta* (p. 207)

6. Branches not biseriate 7

7. Main axes typically flattened, linear-lanceolate, a few slender branches arising from near the base, cells irregular in arrangement in main blades . . . . . . *E. compressa* (p. 208)

7. Not as above; cells usually in regular rows 8

8. Branches mostly restricted to the base; axes 1–2 mm wide . . . . *E. lingulata* (p. 208)

8. Branches typically arising from all levels 9

9. Cell layer 18–28 μm thick; main axes with slender branches but not redivided; a plant of open, exposed places . . . . . .
. . . . *E. erecta* (p. 208)

9. Cell layer 15–20 μm thick; plants of bays, marshes, protected places 10

10. Plants often proliferous along the margins, sparingly branched to second order; cells of main axes to 20 μm in diameter; chromatophores usually covering over ⅔ of cell face . . . . . .
. . . . *E. prolifera* (p. 210)

10. Plants typically repeatedly branched; cells of main axes to 30 μm in diameter; chromatophores usually covering less than ⅔ of cell face . . . . *E. clathrata* (p. 211)

ENTEROMORPHA LINZA (LINNAEUS) J. AGARDH

(Fig. 88)

Plants consisting of a flat, unbranched blade with a hollow, tapering base and a short stalk; the blade linear to lanceolate, green to yellowish-green, 2–3 cm wide, 15–30 cm long when mature, the margin often undulate; cells of the stalk in longitudinal rows, those of the blade in irregular rows in young plants, but in no definite order in older plants, angular in surface view, 10–20 μm in diameter; blades of two adherent

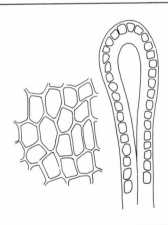

Fig. 88. *Enteromorpha linza:* left, a group of cells in surface view; right, a cross section of a blade at the margin where the two cell layers show separation

cell layers, 35–50 μm thick except at the margins where the cell layers are separated to form a tubular space.

E. *linza* is morphologically intermediate between the genera *Ulva* and *Enteromorpha*. In fact, it was originally named *Ulva linza* by Linnaeus (1753, p. 1163).

On rocks, shells, pilings, in the mid or upper intertidal zone, the plants often in clusters or groups, in bays and estuaries or other protected localities and to some extent in moderately exposed places. This species is best developed in northern waters but is probably continuous from the Florida Keys (in winter) to the Arctic (in summer). From North Carolina southward, it is a plant of winter and spring, appearing and disappearing annually. In the most southern part of its range in the western North Atlantic, it seems to be present only from December through February. In Virginia it may persist through the summer or it may disappear through the warmest months. Its behavior in this respect has not been observed sufficiently. During the entire summer of 1962, it was present in the lower York River area and along the Eastern Shore, but the water temperatures at that time were the lowest in many years, according to records kept at the Virginia Institute of Marine Science.

J. Agardh, 1873–90 (1882), p. 134, pl. 4, figs. 110–12; Hoyt, 1917–18, p. 420; Kylin, 1949, p. 19, fig. 12; Taylor, 1957, p. 68, pl. 3, fig. 8; Mathieson, Dawes, and Humm, 1969, p. 119; Diaz-Piferrer, 1969, p. 163; Zaneveld and Willis, 1974, p. 71, fig. 4; Abbott and Hollenberg, 1976, p. 76, fig. 32; Humm, 1976, p. 85.

ENTEROMORPHA INTESTINALIS (L.) LINK

Plants usually in groups, at first attached, later loose and drifting or entangled, 10–60 cm tall or more, usually unbranched but sometimes proliferating or branched a little at the base. Above the stalk, mature plants are usually inflated and bullate. Cells irregular in arrangement, 9–15 μm in diameter in surface view, the cell layer 20–40 μm thick.

In Virginia this species is widely distributed in the Chesapeake Bay and along the Eastern Shore during winter and spring, though it may persist through the warm months in some places during some years. The distribution and seasonal behavior of E. *intestinalis* seems to be the same as that of E. *linza* (see above). It is known from Florida to the Arctic.

Link, 1820, p. 5; Hoyt, 1917–18, p. 420; Kylin, 1949, p. 22, fig. 13; Taylor, 1957, p. 66, pl. 3, fig. 7, and pl. 4, figs. 4–5; Diaz-Piferrer, 1970, p. 163; Zaneveld and Willis, 1974, p. 70, fig. 3.

ENTEROMORPHA MARGINATA J. AGARDH

Plants small, slender, in tufts 2–3 cm tall, 100–300 μm wide, the width constant except for a tapering at the base and apex; unbranched, or rarely with a few branchlets. Cells in distinct rows at the margins of the flattened blades but irregular in arrangement at the center, 4–8 μm in diameter, rounded or squarish. Chromatophore occupying the entire face of the cell; cell walls relatively thick.

Found in Virginia in Burtons Bay, September 29, 1971, by Dr. F. D. Ott. Previously known from Delaware northward to the Arctic, where it occurs high in the intertidal zone in protected places, especially around the base of salt marsh plants such as *Spartina*.

J. Agardh, 1842, p. 16; J. Agardh, 1883, p. 142; Kützing, 1845–71 (1856), vol. 10, p. 15, pl. 41, fig. 1; Collins, 1909, p. 202; Collins and Hervey, 1917, p. 33; Hamel, 1931, p. 152, fig. 46; Taylor, 1957, p. 65; Zaneveld and Willis, 1974, p. 69.

ENTEROMORPHA MINIMA NÄGELI

Plants small and slender, in dense clusters or bands in the intertidal zone, usually unbranched but sometimes sparingly proliferously branched. Blades linear, light green, soft; 1–10 cm tall, 1–2 mm in diameter, tubular or flattened. Cell layer 8–10 μm thick, the cells 5–7 μm in diameter in surface view, irregularly arranged.

Although primarily a plant of northern waters, this species is known from Biscayne Bay (in brackish water) at Miami, Florida; from Bermuda; and from North Carolina to Newfoundland and Labrador. In Virginia it apparently persists the year around, especially on pilings and seawalls in the upper intertidal zone. From North Carolina southward, it may be a winter-spring species, but adequate observations have not yet been made. Kapraun (1970) did not find it at Port Aransas, Texas, which is surprising since both *Porphyra* and *Petalonia* occur there during winter and spring.

Kylin (1947b) placed this species in the genus *Blidingia*, which he separated from *Enteromorpha* because *Blidingia* arises from a basal disk and produces zoospores only (no gametes). Some phycologists have followed Kylin in this matter (Cardinal, 1965; Villalard, 1967; Lee, 1969; Chapman, 1971; Zaneveld and Willis, 1974).

Nägeli in Kützing, 1849, p. 482; Collins and Hervey, 1917, p. 33; Williams, 1948, p. 685; Taylor, 1957, p. 67; Wilce, 1959, p. 65; Taylor, 1960, p. 62; Mathieson, Dawes, and Humm, 1969, p. 119; Zaneveld and Willis, 1974, p. 71, fig. 5 (as *Blidingia minima*).

ENTEROMORPHA PLUMOSA KÜTZING

Plants small to medium in size, usually scattered, repeatedly branched, the branches delicate, soft, and usually ending in a long, uniseriate tip. Though the branching is mostly irregular, opposite branches are often common. Cells near the ends of the branches 8–12 μm in diameter, in the main axes 12–20 μm in diameter, the chromatophores occupying only part of the faces of the cells.

Common in the Chesapeake Bay on shells, larger algae, salt-marsh grasses; usually below low tide. Widely distributed from the tropical Atlantic northward almost to the Arctic. In the northern part of its range it is a summer species, but in its southern range it is year around.

Bliding (1944) a Swedish authority on the genus *Enteromorpha,* regarded this species as a form of *E. clathrata,* as did Kylin (1949) and Kapraun (1970). Kapraun's work, since it involved laboratory culture as well as field work, is quite convincing, and it may be best to treat this species as a variety of *E. clathrata,* if there are intermediates between the long uniseriate tips of *E. plumosa* and the nonuniseriate or short-uniseriate tips of *E. clathrata,* and if there are intermediates in plant sizes.

Kützing, 1843, p. 300, pl. 20, fig. 1; Collins, 1909, p. 198; Taylor, 1957, p. 63; Mathieson, Dawes, and Humm, 1969, p. 119; Rhodes, 1970, p. 62; Zaneveld and Willis, 1974, p. 68.

ENTEROMORPHA TORTA (MERTENS) REINBOLD

Plants slender, filamentous, often entangled with other algae, sea-grasses, or around the base of salt-marsh plants, or on stones or shells; unbranched or with a few branches from the base, the main axes tubular and bearing long branchlets above that are biseriate or sometimes uniseriate. Cells usually in both longitudinal and transverse rows, the main axes several (4–8) cells wide, the cells about 10 μm in diameter as seen in surface view, isodiametric or a little shorter than wide.

In Virginia this species was found around the base of *Zostera* in the York River in winter by Dr. J. S. Zaneveld. Previously known from Maine northward.

Mertens *in herb.* Juergens, 1816–22 as *Conferva torta*); Collins, 1909, p. 198; Hamel, 1931, p. 154, figs. 46B, 46C; Taylor, 1957, p. 63; Zaneveld and Willis, 1974, p. 66.

ENTEROMORPHA COMPRESSA (L.) GREVILLE

(**Fig. 89**)

Plants in clusters, attached, mostly 8–30 cm tall, the main axes flattened and often resembling *E. linza,* but with a few branches from the base or from the narrow, tapering lower part. Cells small, 10–15 μm in diameter in surface view, irregular in arrangement in the mature parts, somewhat vertically elongate in cross section, the cell layer 13–20 μm thick.

E. compressa is one of the more common members of the genus in the Chesapeake Bay during the summer months, and one of the largest. It is also abundant along the Eastern Shore. It is present the year around. The species is known from the tropical western Atlantic to the Arctic.

Greville, 1830, p. 180, pl. 18; Hamel, 1931, p. 156, fig. 47; Kylin, 1949, p. 22, figs. 14–15; Taylor, 1957, p. 64, pl. 3; Zaneveld and Willis, 1974, p. 69.

ENTEROMORPHA LINGULATA J. AGARDH

Plants to a height of 6–15 cm, in tufts, sparingly to abundantly branched from the lower parts of the main axis, but sometimes unbranched; the larger, better developed branches occurring above the short, slender branchlets. Axes terete, mostly 1–2 mm in diameter, the cells in distinct longitudinal rows, 9–12 μm in diameter, 9–28 μm long, mostly rectangular.

This is a warm-temperate, tropical species previously known from North Carolina to the Caribbean Sea. It was found in Virginia in the York River near Gloucester Point during summer by Dr. Marvin Wass. It is characteristic of more or less exposed places low in the intertidal zone and below, as an epiphyte or on rocks or shells.

J. Agardh, 1883, p. 143; Hamel, 1931, p. 156, fig. 47 (as *E. compressa,* var. *lingulata*); Newton, 1931, p. 69; Blomquist and Humm, 1946, p. 4; Humm and Caylor, 1957, p. 243; Taylor, 1960, p. 60, pl. 1, fig. 3; Humm, 1964, p. 321; Zaneveld and Willis, 1974, p. 69.

ENTEROMORPHA ERECTA (LYNGBYE) J. AGARDH

Plants attached to rocks in exposed localities, soft and slender, 20–40 cm tall, usually with a distinct main axis bearing many smaller, slender

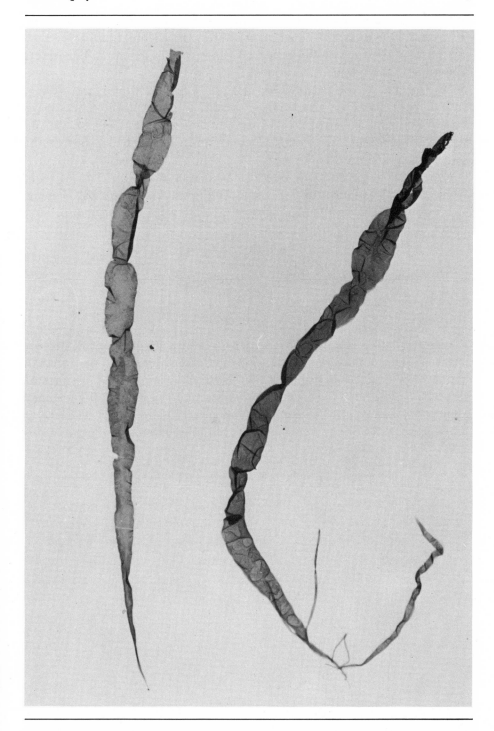

Fig. 89. *Enteromorpha compressa,* photograph of an herbarium specimen showing an unbranched plant and a plant with branching from the base

branches. Cells in distinct longitudinal rows and in transverse rows in the branches, mostly 13–24 μm in diameter, the chromatophore occupying most of the face of the cell. Cell layer 18–28 μm thick. Branchlets sometimes ending in uniseriate tips.

In Virginia on the breakwater at Cape Charles, winter and spring, and at many more or less exposed localities in the lower Chesapeake Bay, including the York River near Yorktown and pilings of the old ferry pier at Gloucester Point. Known from the Caribbean Sea to the Canadian Maritime Provinces.

Lyngbye, 1819, p. 65, pl. 15C (as *Scytosiphon erectus*); J. Agardh, 1883, p. 152; Collins, 1909, p. 200; Williams, 1948, p. 684; Taylor, 1957, p. 64; Wulff and Webb, 1969, p. 31; Diaz-Piferrer, 1970, p. 163; Zaneveld and Barnes, 1974, p. 68.

ENTEROMORPHA PROLIFERA (MÜLLER) J. AGARDH

(Fig. 90)

Plants in tufts or isolated, to a height of 40 cm or more, proliferously branched with occasional branches to the third order. Cells 10–20 μm in diameter, with a chromatophore that usually covers most of the face of the cell; cell layer 15–18 μm thick. The main axes in well-developed plants may be 1–2 cm in diameter and intestiniform. The cells are in longitudinal rows in the younger parts but only vaguely so in the larger axes.

Fig. 90.   *Enteromorpha prolifera,* habit sketch (×¼)

In Virginia this species appears to be present the year around, reaching its best development in late spring. It is widely distributed along the Virginia coastline, especially in protected areas. Known from the Caribbean Sea to James Bay, Canada. In the northern part of its range it is a summer plant; in the tropics it is primarily a winter-spring species.

J. Agardh, 1883, p. 129, pl. 4, figs. 103–4; Collins, 1909, p. 202; Hoyt, 1917–18, p. 419; Hamel, 1931, p. 160, fig. 48 (as *E. compressa,* var. *prolifera*); Taylor, 1957, p. 65, pl. 3, fig. 2; Humm and Taylor, 1961, p. 348, fig. 6B; Kapraun, 1970, p. 214, figs. 15–16, 57–59; Zaneveld and Willis, 1974, p. 69.

ENTEROMORPHA CLATHRATA (ROTH) J. AGARDH

(**Fig. 91**)

Plants at first attached, later often free and entangled or drifting, reaching a height of 30–50 cm, the axes slender and cylindrical, inflated, repeatedly branched. Cells 10–30 μm in diameter, in distinct longitudinal rows except in the main axes of older plants, where the rows become vague. Chromatophore often with three or four pyrenoids, not filling the entire face of the cell.

Common in all marine and brackish waters of Virginia, year around, but best developed during winter and spring. Known from the Caribbean Sea to the Arctic. *E. crinita* (Roth) J. Agardh is considered to be a brackish-water or salt-marsh form of this species (Taylor, 1960).

Fig. 91. *Enteromorpha clathrata,* habit sketch (×¼)

Roth, 1797–1806 (1806), p. 15 (as *Conferva clathrata*) ; J. Agardh, 1883, p. 153; Collins, 1909, p. 199; Kylin, 1949, p. 28, figs. 27–29; Taylor, 1957, p. 63, pl. 3, fig. 1; Humm and Taylor, 1961, p. 343, fig. 6A; Kapraun, 1970, p. 212, figs. 5–6, 60–62; Zaneveld and Willis, 1974, p. 68.

GENUS *MONOSTROMA*

Plants at first saccate, splitting into a flat blade that is one cell thick, the cells uninucleate and with a single platelike chromatophore having one pyrenoid. Intercalary cell division results in blades of considerable size. Reproduction by conversion of vegetative cells into sporangia or gametangia, the zoospores quadriflagellate, the gametes biflagellate, isogamous or anisogamous.

The genus *Monostroma* is treated here sensu Collins (1909) and Taylor (1957). Bliding (1968) divided the genus, retaining in *Monostroma* and the Monostromataceae those species in which the zygotes produce quadriflagellate zoospores directly and those with an isomorphic gametophyte and sporophyte. He placed in the genus *Ulvaria* (Ulvaceae) those with a developmental pattern similar to *Ulva* and *Enteromorpha*. *Monostroma leptodermum* he placed in the genus *Kornmannia*, as there are no pyrenoids. It seems best to follow the older treatment here. More study of Virginia plants is needed, including culture work.

Key to the Species of the Genus *Monostroma*

Cells in surface view with thick walls; blades 20–40 μm thick . . . . *M. oxyspermum*

Cells in surface view with thin walls; blades not over 10 μm thick . . . . *M. leptodermum*

MONOSTROMA OXYSPERMUM (KÜTZING) DOTY
(Figs. 92 and 93)

Plants attached, usually in tufts, intertidal, mostly 3–10 cm tall, 20–40 μm thick. Detached plants in a protected area may become much larger. Cells 7–18 μm or more in diameter, as seen in surface view, the walls between the cells thick and gelatinous.

This species is year around in Virginia. It is common in the intertidal

Fig. 92. *Monostroma oxyspermum,* photograph of an herbarium specimen

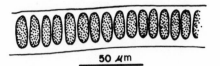

50 μm

Fig. 93.   *Monostroma oxyspermum* cross section

zone of protected areas in bays and salt marshes of the Chesapeake Bay and the Eastern Shore. It is known from Florida to Newfoundland. In the northern part of its range, it is a plant of spring and summer; in the southern part of its range, it is present during winter and spring.

Kützing, 1843, p. 296, pl. 20, fig. 4 (as *Ulva latissima*); Farlow, 1881, p. 42 (as *M. crepidinum*); Collins, 1909, p. 211 (as *M. crepidinum, M. latissimum,* and *M. undulatum,* var. *farlowii*), p. 212 (as *M. orbiculatum,* var. *varium*); Doty, 1947, p. 12; Taylor, 1957, p. 72, pl. 4, figs. 1–3; Humm and Taylor, 1961, p. 352, fig. 6D; Bliding, 1968, p. 585, figs. 31–36 (as *Ulvaria oxysperma*); Mathieson, Dawes, and Humm, 1969, p. 119; Zaneveld and Willis, 1974, p. 72.

MONOSTROMA LEPTODERMUM KJELLMAN

Plants light green, thin and delicate, to about 10 cm tall, rounded-ovate, cuneate, or lanceolate, 7–12 μm thick. Cells quadrangular in surface view, 4–12 μm in diameter, the lateral walls thin, tending to be in longitudinal and transverse rows; in cross section the protoplasts only 4–5 μm high.

This species was found in Virginia by Dr. Barry Wulff on pilings of the old ferry pier at Gloucester Point, where it formed the lowest intertidal band of vegetation during winter and early spring, in association with *Enteromorpha erecta.* It is known from Long Island, New York, to the Arctic Sea.

Whether this species was a temporary member of the flora or whether it occurs each winter and spring has not yet been determined. Other phycologists who have worked in Virginia apparently have not confirmed its presence.

Kjellman, 1877, p. 23, pl. 1; Kjellman, 1883, p. 299; Collins, 1909, p. 213; Wilce, 1959, p. 66; Taylor, 1957, p. 71; Edelstein and McLachlan, 1967b, p. 213, fig. 17; Cardinal, 1967, p. 451, figs. 2C, 3H; Bliding, 1968, p. 606 (as *Kornmannia leptodermum*); Wulff and Webb, 1969, p. 119; Zaneveld and Willis, 1973, p. 72.

GENUS *ULVA*

Plants of expanded sheets, a broad blade, or a basal blade bearing strap-shaped segments or lobes, two cell layers thick. Cells with a single, parietal chromatophore, one to several pyrenoids, uninucleate except for the elongate holdfast extensions of basal cells that become multinucleate. Reproduction by conversion of vegetative cells into sporangia or gametangia, with four to eight quadriflagellate zoospores per cell, or usually eight biflagellate anisogametes per cell. Reproduction is often in patches and may result in holes or perforations in the plants after the walls of the empty reproductive cells have decomposed. Gametophyte and sporophyte isomorphic.

Ulva is one of the four oldest genera of algae, along with *Fucus, Chara,* and *Conferva.* Linnaeus (1753) listed nine species, seven of which are now in other genera. One of these, *Ulva linza* L., is a close relative, having been transferred to *Enteromorpha.* The type specimen of one of the two remaining Linnaean species, *U. latissima,* is believed to be *Laminaria saccharina.* Thus *Ulva lactuca* may be the only surviving Linnaean species of this genus.

Because of the limited number of taxonomically useful and readily visible characters in the genus, it may not be possible to delineate all of the valid species on the basis of anatomy and morphology alone. The name *Ulva lactuca* has been used in all parts of the world for a number of different species in addition to that represented by the type material. Gradually these species are being recognized, as critical studies involving culture work and experimental crossing of gametes are carried on, such as those of Rhyne (1973). His work on what has been called *Ulva lactuca* in North Carolina has shown that species does not occur there, but that there are two *Ulva* species in North Carolina that have not been recognized in the western North Atlantic ocean previously. It seems highly likely that his work applies also to Virginia, as he studied *Ulva* material from Georgia to New Jersey.

Key to the Species of the Genus *Ulva*

Base of plant asymmetric; cells in surface view about 12 μm wide, 12–17 μm long, usually with one pyrenoid . . . . *Ulva curvata*

Base of plant symmetric; cells 12–20 μm wide, 20–30 μm long, usually with two to four pyrenoids . . . . *Ulva rotundata*

ULVA CURVATA (KÜTZING) DE TONI

(**Fig. 94**)

Plants of distromatic, expanded blade with an asymmetric base and stipe when attached, mostly 10–35 cm long but becoming much larger (to one meter or more) when loose and drifting in a favorable environment. Cells about 12 μm in diameter, 12–17 μm long, with a single parietal chloroplast that is usually located along a side wall and with one pyrenoid (rarely two). Plants 35–40 μm thick near the upper margin, about 45 μm thick near the center, 75 μm near the base. Sporophytes producing quadriflagellate zoospores that are 4.5–5.5 μm in

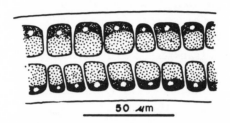

Fig. 94. *Ulva curvata,* cross section

diameter, 8–11 μm long; female plants producing gametes that are 3.5 μm in diameter, 6.4–7.4 μm long; male gametes 2 μm in diameter, 5.4–6.4 μm long. In North Carolina Rhyne (1973) observed that motile cells are released approximately every two weeks, a few days before to one day after each new and full moon. Gametes are capable of developing into gametophyte plants without fertilization, these plants sometimes forming mosaiclike areas of monoploid and diploid cells that later produce biflagellate and quadriflagellate swarmers respectively. The diploid cell areas apparently arise as a result of the failure of chromosome separation during mitosis. Germlings of gametes, zygotes, or zoospores are filamentous and *Enteromorpha*-like at first. It is in this microfilamentous phase that the species is present during the unfavorable (warmest) season of the year.

Ulva curvata was newly reported for North America by Rhyne (1973). He expressed the opinion that it is the only species of *Ulva* occurring in marshes or marshlike habitats between Georgia and New Jersey. It is tolerant of low salinity and salinity fluctuations. It is known in the Mediterranean Sea and the eastern North Atlantic ocean from Morocco to the west coast of Sweden.

Kützing, 1845, p. 245 (as *Phycoseris curvata*); De Toni, 1889, p.

116; Bliding, 1968, p. 570, figs. 23A-I, 24A-J; Rhyne, 1973, p. 18, figs. 2–4, 6, 11–14.

ULVA ROTUNDATA BLIDING

Plants consisting of a flat or undulating blade that is two cell layers thick, with usually a symmetric stipe and base when attached, generally 5–20 cm long but becoming larger when loose and drifting, ovate to orbicular, often lobed, attached to shells, stones, wood, or other solid surfaces in somewhat exposed situations. Blades 50–60 μm thick in the upper half, 75–80 μm thick below; cells in surface view 12–20 μm wide, 20–30 μm long (occasionally larger), with one parietal chloroplast that is usually on the outer face of the cell, pyrenoids two to four (rarely one or five). Zoospores 5.7 μm in diameter, 10 μm long; female gametes 3.6 μm in diameter, 6.7 μm long; male gametes 2.5 μm in diameter, 5 μm long.

Ulva rotundata was newly reported for North America by Rhyne (1973). It occurs in the higher salinity areas of coastal waters, growing best in salinities of 30–35°/oo, although it is tolerant of somewhat lower salinities, at least temporarily. Rhyne's efforts to hybridize U. rotundata and U. curvata were unsuccessful, indicating that they are genetically isolated. The north-south distribution of U. rotundata along the Atlantic coast of North America is not yet known, as it has been confused with U. lactuca. In the eastern North Atlantic it has been recorded from the Canary Islands to northern Norway.

Bliding 1968, p. 566, figs. 20A-D, 21A-H, 22A-D; Rhyne, 1973, p. 22, fig. 5C-E.

ORDER CLADOPHORALES

Plants filamentous, uniseriate, multinucleate, branched or unbranched, the gametophytes and sporophytes isomorphic. Reproduction by biflagellate or quadriflagellate zoospores, by isogametes or anisogametes, or by akinetes in unmodified or slightly swollen vegetative cells that become reproductive. The plants are dioecious.

There is but one family, Cladophoraceae.

GENUS UROSPORA

Plants of uniseriate, unbranched filaments attached by the basal cell and with or without rhizoidal extensions from cells just above the base.

Vegetative cells (coenocytes) short, thick-walled, containing a netlike chromatophore in parietal position. Reproduction by anisogamous, pear-shaped quadriflagellate zoospores and by akinetes.

Members of this genus and of the genus *Codiolum* apparently represent heteromorphic stages of the same plant.

UROSPORA MIRABILIS ARESCHOUG

Plants of uniseriate, unbranched filaments, attached by a basal cell and by rhizoids that arise from the basal cell and from one to three cells above. Rhizoids from the cells above grow down through the thick, gelatinous walls of the cells below. Filaments 2–8 cm long, 20–30 μm in diameter, the cells mostly isodiametric below, but 1–2 diameters long above. Cells that develop into sporangia become 50–55 μm in diameter and are isodiametric; zoospores pyriform, with four flagella at the anterior (blunt) end. Gametes small, anisogamous, with two flagella and an eye spot.

U. mirabilis grows high in the intertidal zone, typically mixed with *Bangia fuscopurpurea* and *Ulothrix flacca,* on rocks in exposed places. In the southern part of its range, it is present during winter and spring. There is a microscopic phase in its life history.

Collected at Cedar Island, Virginia, by Dr. Russell Rhodes, winter of 1968. Previously known from the Faeroe Islands, the British Isles, and the north coast (English Channel) of France.

Areschoug, 1866, p. 15, pl. 3; Børgesen, 1903, p. 500, fig. 100; Hamel, 1931, p. 128, figs. 39 e, 39 f.

GENUS *CHAETOMORPHA*

Plants filamentous, unbranched, attached by a modified basal holdfast cell or occurring as loose, entangled filaments. Cells usually short and often with thick, lamellose walls; filaments typically constricted at the nodes.

Key to the Species of the Genus *Chaetomorpha*

Plants loose, entangled . . . . *C. linum*

Plants attached by a basal holdfast cell . . . . *C. aerea*

CHAETOMORPHA LINUM (MÜLLER) KÜTZING
(Fig. 95)

Filaments unattached, entangled or drifting, cylindrical or slightly constricted at the nodes, 100–375 μm in diameter, the cells mostly 1–2 diameters long, occasionally longer or shorter.

Among other algae on the breakwater at Cape Charles, around the bases of *Spartina* in salt marshes, entangled in *Zostera* at many locali-

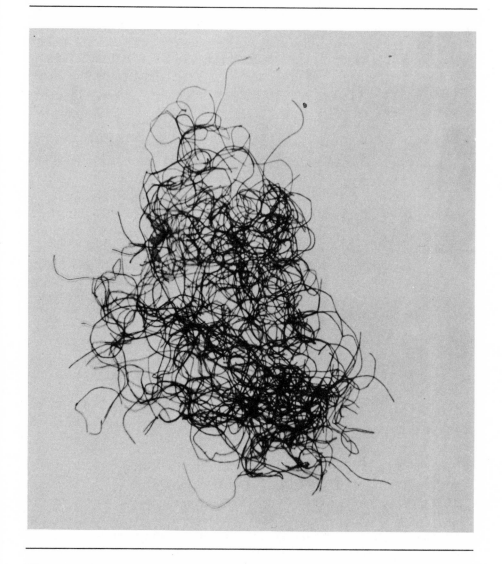

Fig. 95. *Chaetomorpha linum,* photograph of an herbarium specimen

ties in the lower Chesapeake Bay and the Eastern Shore of Virginia. Known from the Caribbean Sea to Newfoundland and Labrador.

Müller 1782, p. 771, fig. 2 (as *Conferva linum*); Kützing, 1845, p. 204; Collins, 1909, p. 325 (as *Chaetomorpha aerea, forma linum*); Hoyt, 1917–18, p. 425; Wood and Palmatier, 1954, p. 140; Taylor, 1957, p. 78, pl. 1, figs. 1–2; Humm and Taylor, 1961, p. 354, fig. 5G; Mathieson, Dawes, and Humm, 1969, p. 120; Zaneveld and Willis, 1974, p. 73.

CHAETOMORPHA AEREA (DILLWYN) KÜTZING

Plants attached by a disklike basal cell, in groups, 10–20 cm tall, 150–350 μm in diameter but more slender near the base. Cells 1–2 diameters long, a little constricted at the nodes; the basal cell 300–900 μm long. Zoospores produced at the upper ends of the filaments, the reproductive cells becoming enlarged and moniliform, 600–700 μm in diameter when mature.

Recorded from Little Creek near Virginia Beach, Virginia, by Zaneveld and Willis (1974) growing on a concrete block in the intertidal zone. Known from the Caribbean Sea to Prince Edward Island.

The morphological similarity and coincidence of range of *C. aerea* and *C. linum* suggest that they may be the attached and unattached forms of the same species, as indicated by Hoyt (1917–18).

Dillwyn, 1802–9, p. 80, pl. 80 (as *Conferva aerea*); Kützing, 1849, p. 379; Farlow, 1881, p. 46; Collins, 1909, p. 324, pl. 12, fig. 115; Hoyt, 1917–18, p. 425, fig. 2A (as *Chaetomorpha linum, forma aerea*); Taylor, 1957, p. 79, pl. 1, figs. 10–12; Zaneveld and Willis, 1974, p. 73.

GENUS *RHIZOCLONIUM*

Filaments prostrate, unbranched or occasionally with short, rhizoidlike branchlets that are of one to a few cells.

The separation of *Chaetomorpha* and *Rhizoclonium* appears to be more a matter of tradition than of taxonomic logic. In general, the cells (coenocytes) of *Chaetomorpha* are short and barrel-shaped and the filaments coarse, while in *Rhizoclonium* the cells are cylindrical and several to many times longer than wide. The rhizoidal branchlets are often absent in *Rhizoclonium*.

Key to the Species of the Genus *Rhizoclonium*

1. Filaments 9–12 μm in diameter 2

1. Filaments 15 μm in diameter or more 3

2. Cells mostly 1–2 diameters long . . . . . . *R. kochianum*

2. Cells 3–7 diameters long . . . . *R. kerneri*

3. Filaments 15–30 μm in diameter . . . . . . *R. riparium*

3. Filaments 40–70 μm in diameter . . . . . . *R. tortuosum*

RHIZOCLONIUM KOCHIANUM KÜTZING

(Fig. 96)

Forming patches of fine, unbranched filaments, often in the intertidal zone, mostly 9–12 μm in diameter, with occasional swollen cells to 18

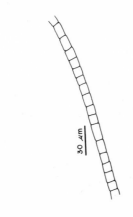

Fig. 96. *Rhizoclonium kochianum*

μm in diameter or more. Cells cylindrical, 1–2 diameters long, with a netlike chromatophore extending the length of the cell, parietal in position. Zoospores about 4 μm in diameter, four (?) to a sporangium.

In Virginia, forming bright green patches in the intertidal zone on pilings, especially where shaded, around the bases of *Spartina* in salt marshes, on sponges and other substrata below low tide. Known from the Caribbean Sea to New England.

Kützing, 1845, p. 206; Collins, 1909, p. 329; Børgesen, 1913, p. 19, fig. 7; Kylin, 1949, p. 50; Humm and Taylor, 1961, p. 358, fig. 5H.

RHIZOCLONIUM KERNERI STOCKMAYER

(Fig. 97)

Filaments unbranched, in patches, entangled, yellowish-green, the cells 10–14 μm in diameter, 3–7 diameters long; plants usually intertidal.

In Virginia it was found on pilings of the old ferry pier at Gloucester Point by Dr. Barry Wulff. It is known from Florida to Newfoundland. It was recorded from Delaware in a tide pool of a mud flat by Dr. J. S. Zaneveld.

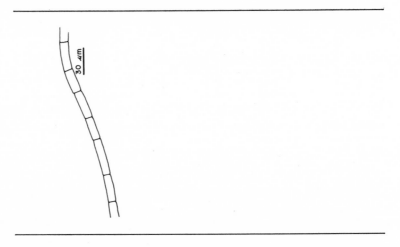

Fig. 97.   *Rhizoclonium kerneri*

Stockmayer, 1890, p. 582; Collins, 1909, p. 329; Børgesen, 1913, p. 20, fig. 8; Humm and Taylor, 1961, p. 359 (as *R. kochianum*, var. *kerneri*); Mathieson, Dawes, and Humm, 1969, p. 120; Zaneveld and Willis, 1974, p. 73.

RHIZOCLONIUM RIPARIUM (ROTH) HARVEY

(Fig. 98)

Plants usually in the form of entangled skeins of filaments, with rhizoidal branches numerous, rare, or lacking. Cells 15–30 μm in diameter, 1–2 diameters long; filaments not constricted at the septa, or only slightly so.

Common in salt marshes, muddy sand flats, and in *Zostera* beds in the lower Chesapeake Bay and along the Eastern Shore of Virginia. Known from the Caribbean Sea to Newfoundland and Labrador, in marine and brackish water.

Harvey, 1846–51, synopsis, p. 314, pl. 238; Collins, 1909, p. 327;

Fig. 98. *Rhizoclonium riparium,* showing the holdfast cell and a rhizoidal branch

Howe, 1920, p. 600; Kylin, 1949, p. 50, fig. 51; Taylor, 1957, p. 81, pl. 1, fig. 3; Wilce, 1959, p. 66; Humm and Taylor, 1961, p. 359, fig. 5J; Mathieson, Dawes, and Humm, 1969, p. 121; Zaneveld and Willis, 1974, p. 73.

RHIZOCLONIUM TORTUOSUM KÜTZING

(Fig. 99)

Plants of dark green, entangled masses of filaments 40–70 μm in diameter, the cells 1–2 diameters long. Rhizoidal branches apparently not produced except in *forma polyrhizum* Holden.

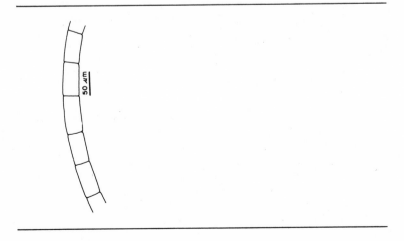

Fig. 99. *Rhizoclonium tortuosum*

In Virginia the species has been found around the base of *Spartina* in salt marshes beside Hummock Channel near Wachapreague and in the James River, summer. Known from Louisiana, Bermuda, and North Carolina to Newfoundland. It is often attached to stones, shells, or woodwork.

Kützing, 1845, p. 206; Collins, 1909, p. 328; Taylor, 1957, p. 80; Mathieson, Dawes, and Humm, 1969, p. 121; Zaneveld and Barnes, 1974, p. 74 (as *Lola capillaris*); Mueller, 1976, p. 92.

GENUS *CLADOPHORA*

Plants of branched, uniseriate filaments, the cells multinucleate and containing a reticulate chromatophore or many small disks, with several pyrenoids. Growth by apical or intercalary cell divisions, or both. Since the "cells" of *Cladophora* are coenocytes, cell division and mitosis are independent processes, although many nuclear divisions may immediately precede formation of a cross wall. Plants attached to the substratum by a holdfast, by rhizoids arising from the basal cell, or by many rhizoidal groups from a basal system of filaments. Reproduction may be strictly asexual or both asexual and sexual. Zoospores are biflagellate or quadriflagellate; gametes are biflagellate. Some species may reproduce by vegetative fragmentation only and be permanently loose and entangled or drifting.

C. van den Hoek (1963) has produced a revision of the European species of the genus. The treatment in this work endeavors to follow Dr. van den Hoek as far as interpretations will permit.

Since species of *Cladophora* other than those treated here may occur in Virginia, this key and the descriptions that follow should not be relied upon entirely. It is suggested that keys in Collins (1909), Taylor (1957), and in van den Hoek (1963) be consulted. Since the latter deals only with European species, it may be necessary to follow the treatment of older American works in many cases.

Key to the Species of the Genus *Cladophora*

1. Plants loose; in salt marshes, quiet water; main axes 100–150 μm in diameter . . . . *C. expansa*

1. Plants typically attached 2

2. Low matted tufts from creeping basal system of filaments; intertidal . . . . *C. albida*

2. Plants not in matted tufts 3

3.   Main filaments mostly under 150 μm in diameter   4

3.   Main filaments mostly over 150 μm in diameter   5

4.   Branchlets and apical cells tapering   .   .   .   *C. sericea* (p. 226)

4.   Branchlets and apical cells not distinctly tapering   .   .   .   .   .
     .   .   .   .   *C. crystallina* (p. 227)

5.   Ultimate branchlets in fasciculate clusters; cell walls less than 5 μm
     thick; main axes without intercalary growth; plants often 15–30
     cm tall or more   .   .   .   .   *C. laetevirens* (p. 227)

5.   Ultimate branchlets not fasciculate, often secund; cell walls 5–10
     μm thick; main axes with intercalary growth; plants usually less
     than 15 cm tall   .   .   .   .   *C. rupestris* (p. 229)

CLADOPHORA EXPANSA (MERTENS) KÜTZING

Plants forming soft, entangled masses or loose cushions, repeatedly
loosely branched. Main axes long, angled-flexuous, 80–150 μm in diameter, the cells 3–6 diameters long or more. Branchlets in part secund,
the ends rounded, about 40 μm in diameter.

In Virginia, entangled in *Zostera* in Willoughby Bay, Norfolk, and in
a protected area at Smith Island, Northampton County. Previously
known from New Jersey to Nova Scotia and Newfoundland. This is a
plant of quiet waters. It is probably a loose form of a common, attached species.

Mertens in herb. Juergens, 1816–22, p. 8 (as *Conferva expansa*);
Kützing, 1845–71 (1854), p. 99, pl. 4; Collins, 1909, p. 343; Taylor,
1957, p. 85, pl. 5, fig. 5; Zaneveld, 1972, p. 129; Zaneveld and Willis,
1974, p. 75.

CLADOPHORA ALBIDA (HUDSON) KÜTZING

Forming mats or spongy tufts of dark green filaments in which the main
axes are pseudodichotomously branched and the growth is principally
intercalary. Main axes 25–100 μm in diameter, the cells 2–4 diameters
long; the branchlets mostly 10–20 μm in diameter. A eurythermal and
euryhaline species that varies considerably with the environment.

Among *Spartina* in the intertidal zone along the shore of the lower
York River near Gloucester Point, Virginia; on rocks below low tide at
Little Creek, near Virginia Beach. Probably widely distributed in the

lower Chesapeake Bay and along the Eastern Shore in moderately protected areas. Previously known from Connecticut to Ile St. Pierre, as
interpreted by American phycologists.

Plants from Virginia that were determined on the basis of Collins
(1909) as *C. magdalenae* Harvey and *C. refracta* (Roth) Areschoug
are included here.

Hudson, 1778, p. 595 (as *Conferva albida*); Kützing, 1843, p. 267;
Harvey, 1846–51, pl. 355a (as *C. magdalenae*); Harvey, 1849, p. 196
(as *C. glaucescens*); Collins, 1909, p. 336; Newton, 1931, p. 86; Taylor, 1957, p. 83, p. 84 (as *C. glaucescens*), p. 88 (as *C. magdalenae*);
van den Hoek, 1963, p. 94, figs. 258–64, 272–76; Zaneveld and Willis,
1974, p. 74 (including *C. glaucescens*), and p. 76 (as *C. magdalenae*).

CLADOPHORA SERICEA (HUDSON) KÜTZING

(Fig. 100)

Plants erect, the main axes often angular or flexuous, 140–160 μm in
diameter in large specimens, pseudodichotomously branched. Ultimate
branchlets often in secund series with cells 3–5 diameters long, the
apical cells 20–60 μm in diameter, the branchlets and apical cells tapering. Growth is principally intercalary.

A variable species with many ecophenes and synonyms. Included here
are plants from Virginia determined *sensu* Collins (1909) as *C. gracilis*
(Griffiths) Kützing, and *C. flexuosa* (Griffiths) Harvey.

From the breakwater at Cape Charles, on shells and adrift at Lynn-

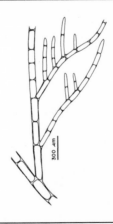

Fig. 100. *Cladophora sericea,* a portion of a plant showing branching

haven Inlet near Virginia Beach, along salt marshes of the York River near Gloucester Point, and from other localities in Virginia.

The species, *sensu* Collins (1909) and of older European treatments of the genus, is known from the Caribbean Sea to Newfoundland.

Hudson, 1762, p. 485 (as *Conferva sericea*); Kützing, 1843, p. 264; Collins, 1909, p. 342 (*C. gracilis*), p. 339 (*C. flexuosa*); van den Hoek, 1963, p. 77, pl. 17, figs. 184–89, pl. 18, figs. 190–209, pl. 19, figs. 210–26, pl. 20, figs. 227–40, and pl. 21, fig. 245.

CLADOPHORA CRYSTALLINA (ROTH) KÜTZING

Plants 10–30 cm tall, light yellow-green, soft and silky, the branching dichotomous to trichotomous, sometimes whorled above. Main axes 80–140 μm in diameter, the branchlets 25–40 μm in diameter, the cells 4–12 diameters long, cylindrical or slightly constricted at the nodes.

In Virginia, from Parramore Island near Wachapreague on shells in a large tide pool; on rocks in Willoughby Bay; summer. Previously known from the Caribbean Sea to Cape Cod.

Roth, 1797–1806, p. 196 (as *Conferva crystallina*); Kützing, 1845, p. 213; Collins, 1909, p. 342; Hoyt, 1917–18, p. 428, pl. 84, fig. 1; Zaneveld, 1972, p. 129; Zaneveld and Willis, 1974, p. 75.

CLADOPHORA LAETEVIRENS (DILLWYN) KÜTZING

(Fig. 101)

Plants to a height of 15–30 cm or more, bushy, with rather wiry main axes, alternately to pseudodichotomously branched, the branches bearing numerous fasciculate clusters of ultimate branchlets. Main axes 200–350 μm in diameter, the cells mostly 2–5 diameters long; apical cells mostly 70–120 μm in diameter, rounded, cylindrical, or with only slight tapering. Intercalary growth occurs in the branches but not in the main axes.

In Virginia, this species is found in many rather exposed situations on rocks, breakwaters, stones, and large shells in the lower Chesapeake Bay and along the Eastern Shore.

This species has been referred to *Cladophora fascicularis* (Mertens) Kützing in the following publications: Humm and Caylor, 1957, p. 244, pl. 5, fig. 5; Taylor, 1960, p. 91, pl. 3, fig. 3; Humm and Taylor, 1961, p. 355, fig. 6F. Interpreted as such, it is a tropical species ranging from the Caribbean Sea to North Carolina and Bermuda.

Fig. 101. *Cladophora laetevirens:* upper left, sketch showing branches in fasciculate groups; below, photograph of an herbarium specimen ($\times \frac{1}{2}$)

Dr. C. van den Hoek (personal communication) has expressed the opinion that this plant is *C. laetevirens*. His descriptions of *C. laetevirens* and *C. vagabunda* (van den Hoek, 1963), however, are almost identical. The only significant differences are the cylindrical apical cells in *C. laetevirens* and the slightly tapering apical cells in *C. vagabunda;* the absence of intercalary growth in the main axes of *C. laetevirens* and the occurrence of some intercalary growth in the main axes of *C. vagabunda*. The distribution of these two species in Europe is essentially the same (Mediterranean Sea to the North Sea). If these two species are distinct, then it seems likely that both could be represented in the Virginia flora also, although only *C. laetevirens* has been recognized thus far.

Dillwyn, 1809, pl. 48 (as *Conferva laetevirens*); Kützing, 1843, p. 267; Collins, 1909, p. 345; Taylor, 1957, p. 87; van den Hoek, 1963, p. 128, pl. 30, figs. 409–17, pl. 31, figs. 418–24, pl. 32, figs. 425–29, and pl. 33, figs. 433, 440; Cardinal, 1967, p. 466.

CLADOPHORA RUPESTRIS (L.) KÜTZING

Plants attached, tufted, dark green, densely branched, stiff, 5–20 cm tall; main axes 90–200 μm in diameter or a little more, the cells 2–7 diameters long. Apical cells 40–80 μm in diameter with thick cell walls, usually 5–10 μm. The dark green color of this plant and the rigidity produced by its thick walls are distinguishing features.

Taylor (1957, p. 88) has expressed some doubt about records of this plant south of Cape Cod. Van den Hoek (1963), however, has recorded it from many stations in the Mediterranean and Adriatic seas, indicating its tolerance of warm-temperate waters.

In Virginia, dwarf plants were found on a plankton buoy in the York River near Gloucester Point in December. It may be limited to winter months in Virginia, but is to be expected along the Eastern Shore. Previously known from Long Island to Nova Scotia and Newfoundland.

Linnaeus, 1753, p. 1167 (as *Conferva rupestris*); Kützing, 1843, p. 270; Farlow, 1881, p. 51 Collins, 1909, p. 346; Taylor, 1957, p. 88, pl. 5, fig. 1; van den Hoek, 1963, p. 64, pl. 15, figs. 146–63; Mathieson, Dawes, and Humm, 1969, p. 120.

ORDER SIPHONALES

Plants completely coenocytic, except that reproductive structures or parts may be separated by a cross wall; filamentous (as in *Bryopsis*), or complex-filamentous in which the filaments are aggregated or interwoven to form plants of various sizes and shapes; or not filamentous, the plants forming more massive coenocytes. Reproduction by zoospores produced in sporangia, by anisogametes or oogamy, by aplanospores, akinetes, or vegetative fragmentation. Nearly all are marine and tropical.

The genera *Ostreobium* and *Vaucheria*, treated in the older literature as members of this order because of their resemblance to Chlorophyta, are placed in the Xanthophyta, primarily because of differences of pigments, stored food, and cell-wall chemistry.

FAMILY BRYOPSIDACEAE

Plants filamentous, erect, bushy, the branchlets distichous or secund. Gametes produced by meiotic nuclear division in the branchlets following formation of a cross wall, or in special gametangia.

Two species of the genus *Bryopsis* are the most eurythermal members of the order Siphonales in the North Atlantic.

GENUS *BRYOPSIS*

Plants filamentous, densely branched, the upper branches developing into gametangia following formation of a cross wall. The plants are diploid and dioecious, the microgametes brownish, the megagametes greenish.

Key to the Species of the Genus *Bryopsis*

Ultimate branchlets distichous, forming fernlike blades . . . . . .
. . . . *B. plumosa*
Ultimate branchlets radial in origin . . . . . . *B. hypnoides*

BRYOPSIS PLUMOSA (HUDSON) C. AGARDH

(Fig. 102)

Plants 6–12 cm tall, densely branched, the branching generally pyramidal in form, the main axes often becoming denuded below. Branchlets distinctly more slender than the branches that bear them and much constricted at the base, 65–100 μm in diameter, distichously arranged producing more or less triangular fernlike blades.

Fig. 102. *Bryopsis plumosa,* photograph of an herbarium specimen (×2)

In Virginia, widely distributed throughout the lower Chesapeake Bay and along the Eastern Shore on breakwaters, rocks, shells, pilings; best developed in winter and spring, but present throughout the year. Known from South Carolina to Nova Scotia and the Bay of Fundy. In North and South Carolina, this species is present only during the colder months, appearing annually in November and disappearing in April or May. It is probably present during the warm months in the form of a dormant holdfast.

Records of this species from Florida, the Caribbean Sea, Gulf of Mexico, and other parts of the tropical Atlantic Ocean are open to doubt. These records may be based upon another species that is morphologically very similar to *B. plumosa,* but sufficiently different in physiological characteristics to justify species separation.

Hudson, 1778, p. 571 (as *Ulva plumosa*); C. Agardh, 1820–28, p. 448; Farlow, 1881, p. 59, pl. 4, fig. 1; Collins, 1909, p. 403, pl. 17, fig. 155; Hamel, 1931, p. 61, fig. 20C; Kylin, 1949, p. 66, fig. 64; Taylor, 1957, p. 94, pl. 7, figs. 1–3; Zaneveld and Willis, 1974, p. 76.

BRYOPSIS HYPNOIDES LAMOUROUX

(Fig. 103)

Plants erect, in dense tufts 5–10 cm tall, dark green, the erect branches considerably branched in an irregular manner to progressively smaller branchlets that are not notably different and not distinctly distichous or plumose. The ultimate branchlets are in radial arrangement.

In Virginia this species is widely distributed in the lower Chesapeake Bay and along the Eastern Shore. It was first reported by Dr. Charlotte Mangum from Sandy Point in the York River near Yorktown growing on the tubes of an annelid worm, *Diopatra*. It is known from the Caribbean Sea, the Gulf of Mexico, Bermuda, and from Florida to Nova Scotia.

Lamouroux, 1809, p. 135, pl. 1, figs. 2a-b; Vickers and Shaw, 1908, p. 30, pl. 53; Collins, 1909, p. 403; Hamel, 1931, p. 68, fig. 20B; Taylor, 1957, p. 94; Edelstein and McLachlan, 1967*b*, p. 211, fig. 19; Edwards, 1970, p. 22, figs. 51–52; Zaneveld, 1972, p. 129, fig. 3; Zaneveld and Willis, 1974, p. 76, fig. 7.

Fig. 103. *Bryopsis hypnoides*, photograph of an herbarium specimen (×2)

Glossary
Literature Cited
Index

GLOSSARY

Acuminate: Long-tapering to a slender point

Acute: Tapering to a point, but not long-tapering

Adventitious: Produced in an abnormal or unusual place

Akinete: A spore formed by transformation of a vegetative cell in which food reserves and pigments accumulate and the spore wall is fused with the old wall of the vegetative cell

Anastomosing: Having occasional connections among the parts of a branch or filament system to form a network

Apiculate: Ending abruptly in a short, sharp tip

Arcuate: Curved like a bow

Attenuation: Narrowing upward; having a tapering tip

Axial: Located at the longitudinal center of an elongate part

Bullate: Having bumps or protrusions that are irregular in size and arrangement

Calcified: Having accumulated calcium carbonate, and usually hardened

Chromatophore: A colored plastid within a cell that absorbs light in photosynthesis

Coccoid: Tending to be spherical

Coenocyte (*adj.,* coenocytic): A multinucleate part of a plant, or an entire plant, in which there are no internal walls

Conceptacle: a cavity opening at the surface of an alga within which reproductive cells are produced

Conical: Cone-shaped

Contiguous: Adjacent, connected to

Cortex: In an axis or branch of an alga, the tissue surrounding the central cylinder or medulla

Cortication: The tissue formed secondarily that covers the central cylinder

Cryptostomata: Pores or openings of small internal cavities in the Fucales, believed to be vestigial conceptacles

Deciduous: Falling off naturally

Distichous: Having the parts borne in two vertical rows or ranks on opposite sides

Endophytic: Growing within the tissues of a plant

Endospore: One of many small cells formed by repeated division of a large cell in the family Chamaesiphonaceae of the bluegreen algae; endospores may eventually escape from the parent cell sheath.

Endozooic: Growing within the tissue of an animal

Epiphytic: Growing upon a plant

False branch: In bluegreen algae, the result of the branching of the sheath of a filamentous species, but not involving the branching of a trichome

Fascicle: A cluster of parallel filaments or branches

Fasciculate: Tending to be crowded into bundles in which the units are parallel

Filament: A row of cells attached end to end; in bluegreen algae, the term includes both sheath and trichome.

Gametangium: A cell or structure in which gametes are produced

Gland cell: A small, superficial cell or a cell buried within the tissue and differing from a normal vegetative cell; it is usually translucent and without granules or distinct cell inclusions.

Gonidia: An old term used to refer to the endospores of bluegreen algae

Heterocyst: In red algae of the family Corallinaceae, an enlarged color-less cell; in bluegreen algae, a thick-walled, yellowish-translucent cell transformed from a vegetative cell and having a gelatinous plug at one or both ends

Hyaline: Transparent and colorless

Intercalary: Between apex and base, not terminal

Lanceolate: Lance-shaped; broadest at a place below the center and tapering to the apex

Lateral: Originating at the side

Linear: Much elongated and with parallel margins for most of its length

Monosporangia: Sporangia that produce a single, asexual spore; char-acteristic of the genus *Acrochaetium*

Multinucleate: Having more than one nucleus; coenocytic

Multiseriate: Composed of more than one row of cells

Node: In an axis, the plane of origin of a branch or appendage; in a filament, the plane of a cross wall

Nodulose: Having small knobs or bumps

Ovoid: Tending to be oval in shape

Paraphyses: Sterile, branched or unbranched filaments of determinate growth that occur among reproductive structures

Parenchymatous: Consisting of a tisue of thin-walled nonfilamentous cells that are more or less isodiametric

Parietal: Located just inside the wall of a cell; lining the wall

Pelagic: Occurring in the open sea and floating or in suspension in the water

Pericentral cells: A ring of cells or filaments of cells that surround a central cell or filament

Physodes: Small vesicles in the cells of all or most brown algae, often grouped around the nucleus, and containing phloroglucinol derivatives and tannins; formerly referred to as fucosan vesicles

Pinnate: Branching that resembles a feather; distichous

Planktonic: Living in suspension in the water and lacking sufficient motility to overcome currents

Plurilocular: Multicellular

Pluriseriate: Consisting of more than a single row of cells, at least in part

Polysiphonous: Characteristic of a filament or branch composed of more than a single row of cells, or "siphons"

Proliferous: Branching produced late in the development of a plant or an axis, sometimes in response to injury or grazing

Protoplast: The contents of a cell inside the wall; the living material

Pseudoparenchymatous: Having a cell structure that resembles parenchyma but is of filamentous origin

Pseudovacuole: A gas-filled, vacuolelike area in the cell of a bluegreen alga that develops in response to certain conditions

Pyriform: Pear-shaped; wider at the bottom and tapering upwards

Receptacle: The enlarged reproductive part of a plant, or special branchlets, in which conceptacles are produced

Rhizoid: A cellular filamentous outgrowth serving as an organ of attachment, often colorless

Secund: Said of branchlets or appendages that arise in a single row along an axis; comblike

Seriate: In a row

Sheath: A polysaccharide outer coating secreted by most bluegreen algae and, in many filamentous genera, loose so that the trichome can move out of it

Sorus: A group of reproductive organs (plural, sori)

Sporangium: A structure in which spores are borne

Stellate: Star-shaped

Terete: Cylindrical; round, as viewed in cross section

Tetrasporangia: Sporangia that produce four spores only

Torulose: Twisted, knobby

Trichoblast: Colorless, branched or unbranched filament produced at the growing tip of some red algae, and soon deciduous

Trichome: In filamentous bluegreen algae, the row of cells not including the sheath

Truncate: Blunt at the end; more or less squared off

Unilocular: Consisting of a single cell

Uniseriate: Consisting of a single row of cells

Vesicle: A hollow, gas-filled, bladderlike structure

Whorled: Arising in a ring around an axis

LITERATURE CITED

Abbott, Isabella A., and G. J. Hollenberg. 1976. Marine Algae of California. Stanford, Calif. xii plus 827 pp., 701 figs.

Agardh, C. A. 1810–12. Disposito algarum Sueciae. Pts. 1–4. Lund. 45 pp.

————. 1817. Synopsis algarum Scandinaviae, adjecta dispositione universali algarum. Lund. xl plus 135 pp.

————. 1820–28. Species algarum rite cognitae cum synonymis, differentiis specificis et descriptionibus succinctis. Lund. Vol. 1, pt. 1, pp. 1–168, 1820; vol. 1, pt. 2, pp. 169–531, 1822; vol. 2, pt. 1, pp. i–lxxvi plus 1–189, 1828.

————. 1824. Systema algarum. Lund. xxxviii plus 312 pp.

————. 1827. Aufzählung einiger in den östreichischen Ländern gefundenen neuen Gattungen und Arten von Algen, nebst ihrer Diagnostik und beigefugten Bemerkungen. Flora 10:625–46.

Agardh, J. G. 1841. In historiam algarum symbolae. Linnaea 15:1–50, 443–57.

————. 1842. Algae Maris Mediterranei et Adriatici. Paris, x plus 164 pp.

————. 1848–76. Species, genera, et ordines algarum. Lund. Vol. 1, viii plus 363 pp., 1848; vol. 2, xii plus 351 pp., 1851; vol. 3, vii plus 724 pp., 1876.

————. 1873–90. Till Algernes Systematik. I-III. Lunds Univ. Arsskrift 9:1–71, 1873; IV-V. Ibid. 17:1–134, 1880; VI. Ibid. 19:1–177, 1882; VII. Ibid. 21:1–117, 1884; VIII. Ibid. 23:1–174, 1887; IX-XI. Ibid. 26:1–125, 1890.

————. 1892. Analecta algologica. Lunds Univ. Arsskrift 28:1–182.

Areschoug, J. E. 1843. Algarum (Phycearum) minus rite cognitarum pugillus secundus. Linnaea 17:257–269.

————. 1846–50. Phyceae Scandinavicae marinae. Uppsala.

————. 1866. Observationes phycologicae. Pt. 1. De confervaceis non-nullis. Uppsala. 26 pp., 4 pls.

Aziz, K. M. S. 1965. *Acrochaetium* and *Kylinia* in the southwestern North Atlantic Ocean. Ph.D. thesis, Duke University, 235 pp., 16 pl.

————, and H. J. Humm. 1962. Additions to the algal flora of Beaufort, N.C., and vicinity. Jour. Elisha Mitchell Sci. Soc. 78:55–63.

Bailey, J. W. 1848. Notes on the algae of the United States, part 2. Amer. Jour. Sci. and Arts, ser. ii, 6:37–42.

Baker, A. F., and H. C. Bold. 1970. Phycological studies. X. Taxonomic studies in the Oscillatoriaceae. Univ. of Texas Publ. no. 7004, Austin. 105 pp., 125 figs.

Ballantine, D., and H. J. Humm. 1975. Benthic algae of the Anclote estuary. I. Epiphytes of seagrass leaves. Florida Sci. 38:150–62.

Batters, E. A. L. 1902. A catalogue of the British marine algae. Jour. Bot. 40 (supp.):1–107.

Bird, N., J. McLachlan, and D. Grund. 1977. Studies on *Gracilaria*. 5. *In vitro* life history of *Gracilaria sp.* from the maritime provinces. Canadian Jour. Bot. 55:1282–90.

Blackwelder, Brenda C. 1975. Attached forms of intertidal algae from the coastal regions of South Carolina. Ph.D. thesis, University of South Carolina. 322 pp., 117 figs., 46 tables.

Bliding, C. 1944. Zur Systematik der schwedischen *Enteromorpha*. Bot. Notiser, 1944, pp. 331–56.

————. 1968. A critical survey of European taxa in Ulvales. II. *Ulva, Ulvaria, Monostroma, Kornmannia*. Bot. Notiser 121:535–629.

Blomquist, H. L. 1954. A new species of *Myriotrichia* from the coast of North Carolina. Jour. Elisha Mitchell Sci. Soc. 70:37–41.

————. 1958. *Myriotrichia scutata* Blomquist conspecific with *Ectocarpus subcorymbosus* Holden in Collins. Jour. Elisha Mitchell Sci. Soc. 74:24.

————, and H. J. Humm. 1946. Some marine algae new to Beaufort, N.C. Jour. Elisha Mitchell Sci. Soc. 62:1–8.

Børgessen, F. 1903. The marine algae of the Faeroes *in* Botany of the Faeroes Based upon Danish Investigations, vol. 2, pp. 339–532. Copenhagen and Christiania.

————. 1913. Marine algae of the Danish West Indies. I. Chlorophyceae. Dansk Bot. Arkiv 1(4), 158 pp., 126 figs., 1 map.

————. 1914. Marine algae of the Danish West Indies. II. Phaeophyceae. Dansk Bot. Arkiv. 2(2), 66 pp., 44 figs.

————. 1916–20. Marine algae of the Danish West Indies. III. Rhodophyceae. Dansk Bot. Arkiv 3(1a–1f), 498 pp. 435 figs.

————. 1932. A revision of Forsskal's algae mentioned in Flora Aegyptiaco-Arabica and found in his herbarium in the Botanical Museum of the University of Copenhagen. Dansk Bot. Arkiv 8(2), 15 pp., 1 pl.

Bornet, E. 1904. Deux *Chantransia corymbifera* Thuret. Bull. Soc. Bot. de France (session extraord.) 51:14–23.

————, and C. Flahault. 1886–88. Revision des Nostocacees Heterocystees contenues dans les principaux herbiers de France, I–IV. Ann. Sci. Nat., Bot., ser. 7, 3:323–81; 4:343–73; 5:51–129; 7:177–262.

————, and ————. 1888. Note sur deux noveaux genres d'algues perforantes. Jour. Bot. 2:161–65.

————, and ————. 1889. Sur quelques plantes vivant dans le test calcaire des mollusques. Bull. Soc. Bot. de France 36:cxlvii–clxxvi.

Bory de Saint Vincent, J. B. 1822. *In* Dictionaire classique d'histoire naturelle. Vol. 1. Paris. *Anabaina,* p. 307; Arthrodiees, pp. 591–98.

Breed, R. S., R. G. E. Murray, and N. R. Smith (editors). 1957. Bergey's Manual of Determinative Bacteriology, 7th ed. Baltimore. 1132 pp.

Buchanan, R. E., and N. E. Gibbons (editors). 1974. Bergey's Manual of Determinative Bacteriology, 8th ed. Baltimore. 1246 pp.

Cardinal, A. 1965. Liste preliminaire des algues benthiques de la Baie des Chaleurs. Rapp. ann. 1964, Sta. Biol. Mar., Grande-Riviere, Quebec, Canada, pp. 41–51.

————. 1967. Inventaires des algues marines de la Baie des Chaleurs et de la baie de Gaspé (Quebec). II. Chlorophyceae. Naturaliste Canadien 94:447–69.

————. 1968. Repertoire des algues marines benthiques de l'est du Canada. Cahiers d'Information no. 48, Sta. Biol. Mar., Grande-Riviere, Quebec, 213 pp.

Carr, N. G., and B. A. Whitton (editors). 1973. The Biology of the Blue-green Algae. Berkeley and Los Angeles. x plus 676 pp.

Causey, Nelle B., J. P. Prytherch, Jean McCaskill, H. J. Humm, and F. A. Wolf. 1946. Influence of environmental factors upon the growth of *Gracilaria confervoides*. Duke Univ. Mar. Sta. Bull. 3:19–24.

Chamberlain, Yvonne M. 1977. The occurrence of *Fosliella limitata* and *F. lejolisii* on the Isle of Wight. Brit. Phycol. Jour. 12:67–81.

Chapman, R. L. 1971. The macroscopic marine algae of Sapelo Island and other sites on the Georgia coast. Bull. Georgia Acad. Sci. 29:77–89.

Chen, L. C-M., T. Edelstein, E. Ogata, and J. McLachlan. 1970. Life history of *Porphyra miniata*. Canadian Jour. Bot. 48:385–89.

Cocke, E. C. 1967. The Myxophyceae of North Carolina. Published by the author, Wake Forest University, Winston-Salem, N.C. vii plus 206 pp., 326 figs.

Coll, J., and J. Cox. 1977. The genus *Porphyra* in the American North

Atlantic. I. New species from North Carolina. Botanica Marina 20: 155–59.

Collins, F. S. 1905. Phycological notes of the late Isaac Holden. Rhodora 7:222–43.

———. 1906. *Acrochaetium* and *Chantransia* in North America. Rhodora 8:189–96.

———. 1909. The green algae of North America. Tufts College Stud. (sci. ser.), 2(3):79–480, 18 pls.

———. 1918. Notes from the Woods Hole Laboratory–1917. Rhodora 20:141–45.

———, and A. B. Hervey. 1917. The algae of Bermuda. Proc. Amer. Acad. Arts and Sci. 53:1–195, 6 pls.

Crouan, P. L., and H. M. Crouan. 1867. Florule du Finistère, contenant les descriptions de 360 èspeces nouvelles de sporogames observationes et une synonymie des plantes cellulaires et vasculares qui croissent spontanément dans ce départment. Paris. x plus 262 pp., 32 pls.

Dawson, E. Y. 1960. Marine red algae of Pacific Mexico. Part 3. Cryptonemiales, Corallinaceae, subfam. Melobesioideae. Pacific Naturalist 2(1), 125 pp. 50 pls.

De la Pylaie, B. 1829. Flore de l'île de Terre Neuve et des iles Saint Pierre et Miclon avec figures dessinées par l'auteur sur la plante vivante, pl. 17. Paris. 128 pp. (The plates were never published.)

DeLoach, W. S., C. Wilton, H. J. Humm, and F. A. Wolf. 1946. Preparation of an agar-like gel from *Hypnea musciformis*. Duke Univ. Mar. Sta. Bull. 3:31–39.

———, ———, J. McCaskill, H. J. Humm, and F. A. Wolf. 1946. *Gracilaria confervoides* as a source of agar. Duke Univ. Mar. Sta. Bull. 3:25–30.

Desikachary, T. V. 1959. Cyanophyta. New Delhi. 686 pp., 137 pls.

De Toni, J. B. 1889. Sylloge algarum omnium hucusque cognitarum. I. (chlorophycearum). Patavii. 1315 pp.

Diaz-Piferrer, M. 1969. Distribution of the benthic flora of the Caribbean Sea. Carib. Jour. Sci. 9:151–78.

———. 1970. Adiciones a la flora marina de Venezuela. Carib. Jour. Sci. 10:159–198.

Dillwyn, L. W. 1802–9. British Confervae, or coloured figures and descriptions of the British plants referred by botanists to the genus *Conferva*. London. 87 pp., 109 pls.

Dixon, P. S., and L. M. Irvine. 1970. Miscellaneous notes on algal taxonomy and nomenclature III. Bot. Notiser 123:474–87.

———, and W. N. Richardson. 1969. Life histories of *Bangia* and *Porphyra* and the photoperiodic control of spore production. Pp. 133–39 in Proc. Sixth Internat. Seaweed Symposium. Madrid.

Doty, M. S. 1947. The marine algae of Oregon. Part I. Chlorophyta and Phaeophyta. Farlowia 3:1–65.

Drouet, F. 1938. Notes on Myxophyceae, I–IV. Bull. Torrey Bot. Club 65:285–92.

————. 1951. Cyanophyta. Chap. 8, pp. 159–66, *in* G. M. Smith, editor, Manual of Phycology. Waltham, Mass.

————. 1968. Revision of the Classification of the Oscillatoriaceae. Monograph 15, Acad. of Nat. Sci., Philadelphia. 370 pp., 131 figs.

————. 1973. Revision of the Nostocaceae with Cylindrical Trichomes. New York. 292 pp., 87 figs.

————. 1978. Revision of the Nostocaceae with Constricted Trichomes. Vaduz, Leichtenstein. 258 pp., 42 figs.

————, and W. A. Daily. 1939. The planktonic fresh water species of *Microcystis*. Field Museum Bot. Ser. 20:67–83.

————, and ————. 1948. Nomenclatural transfers among the coccoid algae. Lloydia 11:77–79.

————, and ————. 1956. Revision of the coccoid Myxophyceae. Butler Univ. Bot. Stud. 12:1–218.

Earle, Linda C., and H. J. Humm. 1964. Intertidal zonation of algae in Beaufort harbor. Jour. Elisha Mitchell Sci. Soc. 80:78–82.

Earle, Sylvia A. 1969. Phaeophyta of the eastern Gulf of Mexico. Phycologia 7(2):71–254.

Edelstein, T., C. J. Bird, and J. McLachlan. 1974. Tetrasporangia and gametangia on the same thallus in the red algae *Cystoclonium purpureum* and *Chondria baileyana*. Brit. Jour. Phycol. 9:247–50.

————, and J. McLachlan. 1966. Investigations of the marine algae of Nova Scotia. I. Winter flora of the Atlantic coast. Canadian Jour. Bot. 44:1035–55.

————, and ————. 1967a. Investigations of the marine algae of Nova Scotia. III. Species of Phaeophyceae new or rare to Nova. Scotia. Canadian Jour. Bot. 45:203–10.

————, and ————. 1967b. Investigations of the marine algae of Nova Scotia. IV. Species of Chlorophyceae new or rare to Nova Scotia. Canadian Jour. Bot. 45:211–14.

————, ————, and J. S. Craigie. 1967. Investigations of the marine algae of Nova Scotia. II. Species of Rhodophyceae new or rare to Nova Scotia. Canadian Jour. Bot. 45:193–202.

Edwards, P. 1969. Life History of *Callithamnion byssoides* in culture. Jour. Phycol. 5:266–68.

————. 1970. Illustrated guide to the seaweeds and sea grasses in the vicinity of Port Aransas, Texas. Contrib. in Mar. Sci., University of Texas, vol. 15 (supp.). Austin. 128 pp., 255 figs.

Farlow, W. G. 1881. The marine algae of New England. Report of the U.S. Commissioner of Fish and Fisheries for 1879, app. A-1. Washington, D.C. 210 pp.

Feldmann-Mazoyer, G. 1940. Recherches sur les Ceramiacées de la Mediterranée Occidentale. Algiers. 510 pp., 191 figs., 4 pls.

Fiore, J. 1972. Life history studies of New England Phaeophyta. Pp. 12–13 *in* M. Carriker (editor), Systematics and Ecology Program, Ninth and Tenth Annual Reports on Progress and Ten Year Summary. S.E.P., Marine Biological Laboratory, Woods Hole, Mass.

―――. 1975. A new generic name for *Farlowiella onusta* (Phaeophyta). Taxon 24:497–98.

―――. 1977. Life history and taxonomy of *Stictyosiphon subsimplex* Holden (Phaeophyta, Dictyosiphonales) and *Farlowiella onusta* (Kützing) Kornmann *in* Kuckuck (Phaeophyta, Ectocarpales). Phycologia 16:301–11.

Forsskål, p. 1775. Flora Aegyptiaco-Arabica, sive descriptiones plantarum quas par Aegyptum inferiorem et Arabiam felicem detexit post mortem auctoris edidit C. Niebuhr. Hauniae. 33 plus cxxvi plus 219 pp., 1 map.

Foslie, M. H. 1909. Algologiske notiser, VI. K. Norske Vidensk. Selsk. Skr. 1909(2):1–63.

―――. 1929. Contributions to a monograph of the Lithothamnia. Trondhjem. 60 pp., 75 pls.

Fremy, P. 1934. Cyanophycées des côtes d'Europe. Mem. Soc. Sci. Nat. et Math. de Chèrbourg 41:1–234, 66 pls.

Fritsch, F. E. 1951. Chrysophyta. Chap. 5, pp. 83–102, *in* G. M. Smith (editor), Manual of Phycology. Waltham, Mass.

Funk, G. 1955. Beiträge zur Kenntnis der Meeresalgen von Neapel. Publ. della Staz. Zool. di Napoli, vol. 25 (supp.), 178 pp., 30 pls.

Garbary, D. J., D. Grund, and J. McLachlan. 1978. The taxonomic status of *Ceramium rubrum* based on culture experiments. Phycologia 17:85–94.

Gomont, M. 1892. Monographie des Oscillariées. Ann. Sci. Nat., Bot., ser. 7, 15:263–68; 16:91–264.

Goodband, S. J. 1971. The taxonomy of *Sphacelaria cirrosa, Sphacelaria fusca,* and *Sphacelaria furcigera,* a simple statistical approach. Ann. Bot. 35:957–80.

Goodenough, S., and T. J. Woodward. 1797. Observations on British Fuci. Trans. Linn. Soc. London 3:84–235.

Greville, R. K. 1823–29. Scottish Cryptogamic Flora, or coloured figures and descriptions of cryptogamic plants belonging chiefly to the order Fungi, and intended to serve as a continuation of English Botany. Vols. 1–6. Edinburgh. Vol. 1, 1823, 60 pls.; vol. 2, 1824, 60 pls.; vol. 3, 1825, 60 pls.; vol. 4, 1826, 60 pls.; vol. 5, 1827, 60 pls.; vol. 6, 1829, 60 pls.

―――. 1830. Algae Britannicae, or descriptions of the marine and other inarticulated plants of the British Islands, belonging to the order, Algae, with plates illustrative of the genera. Edinburgh. i–lxxxviii plus 218 pp., 19 pls.

Halfen, L. N., and R. W. Castenholz. 1971. Gliding motility in a blue-green alga. Jour. Phycol. 7:133–45.

Hamel, G. 1931. Chlorophyceae des cotes francaises. Protococcales. Rev. Algologique 5(1):1–54. Siphonales. 6(1):9–73.

———. 1931–39. Phaeophycees de France. Paris. xlvi plus 432 pp., 60 figs., 10 pls.

Hamm, D., and H. J. Humm. 1976. Benthic algae of the Anclote estuary. II. Bottom-dwelling species. Fla. Sci. 39:209–29.

Harder, R. 1948. Einordnung von *Trailliella intricata* in den Generationswechsel der Bonnemaisoniaceae. Nachr. Acad. Wiss. Göttingen 1948:24–27.

Harvey, W. H. 1833. Algae. *In* W. J. Hooker, The English Flora of Sir James Edward Smith. Class XXIV, Cryptogamia. Vol. 5, pt. 1, comprising the Mosses, Hepaticae, Lichens, Characeae, and Algae. London. x plus 432 pp.

———. 1834. Notice of a collection of algae communicated to Dr. Hooker by the late Mrs. Charles Telfair from "Cap Malheureux" in Mauritius, with descriptions of some new and little known species. *In* W. J. Hooker's Journal of Botany, vol. 1. London. p. 147.

———. 1846–51. Phycologia Britannica. London. Vols. I–IV, 360 pls. pls. 1–72, 1846; pls. 73–144, 1847; pls. 145–216, 1848; pls. 217–258, 1849; pls. 259–354, 1849–51; pls. 355–60, 1851.

———. 1849. A manual of British Marine Algae, Containing Generic and Specific Descriptions of All the Known British Species of Seaweeds, with Plates to Illustrate All the Genera. London.

———. 1851. Nereis Boreali Americana. Pt. 1. Melanospermae. Smithsonian Contrib. to Knowledge 3(4), 150 pp., 12 pls.

———. 1853. Nereis Boreali Americana. Pt. 2. Rhodospermae. Smithsonian Contrib. to Knowledge 5(5), 258 pp., pls. 13–36.

———. 1858. Nereis Boreali Americana. Pt. 3. Chlorospermae. Smithsonian Contrib. to Knowledge, vol. 10, 140 pp., pls. 37–50.

Heydrich, F. 1892. Beiträge zur Kenntnis der Algenflora von Kaiser-Wilhelmsland (Deutsch Neu Guinea). Ber. der Deut. Bot. Ges. 10:458–85, pls. xxiv–xxvi.

Holden, I. 1899. Two new species of marine algae from Bridgeport, Conn. Rhodora 1:197–98.

Hollenberg, G. J., and Isabella A. Abbott. 1966. Supplement to G. M. Smith's Marine Algae of the Monterey Peninsula. Stanford, Calif. 125 pp., 53 figs.

Hooker, W. J. 1833. The English Flora of Sir James Edward Smith.

Class XXIV, Cryptogamia. Vol. 5, pt. 1, comprising the Mosses, Hepaticae, Lichens, Characeae, and Algae. London. x plus 432 pp.

Howe, M. 1914. The marine algae of Peru. Mem. Torrey Bot. Club 15:1–185, 44 figs. 66 pls.

————. 1920. Algae. Pp. 553–626 *in* N. L. Britton and C. F. Mills-paugh, the Bahama Flora. New York.

Hoyt, W. D. 1907. Periodicity in the production of the sexual cells of *Dictyota dichotoma*. Bot. Gaz. 43:383–92.

————. 1917–18. Marine algae of Beaufort, N.C., and adjacent regions. Bull. U.S. Bur. Fish. 36:367–556, 47 figs., 36 pls., 9 tables.

————. 1927. The periodic fruiting of *Dictyota* and its relation to the environment. Amer. Jour. Bot. 14:592–619.

Hudson, G. 1762. Flora Anglica: exhibens plantas per regnum Britanniae sponte crescentes, distributas secundum systema sexuale, cum differentiis specierum, synonymis auctorum, moninibus incolarum, solo locorum, tempore florende, officinalibus pharmacopaeorum. Ed. 1. London. viii plus 506 pp.

————. 1778. Ibid. Ed. 2. London. xxxviii plus 690 pp.

Humm, H. J. 1942. Seaweeds at Beaufort, N.C., as a source of agar. Science 96:230–31.

————. 1944. Agar resources of the south Atlantic and east Gulf coasts of the U.S. Science 100:209–12.

————. 1948. Method of extracting gelose from seaweeds such as *Hypnea musciformis* and other species of the genus *Hypnea*. U.S. Patent 2,446,091, July 27.

————. 1951a. The seaweed resources of North Carolina. Pp. 231–50 *in* H. F. Taylor (editor), Survey of the Marine Fisheries of North Carolina. Chapel Hill, N.C.

————. 1951b. Agar and related phycocolloids. Chap. 5 *in* D. K. Tressler and J. M. Lemon, Marine Products of Commerce, 2nd ed. New York.

————. 1952. Notes on the marine algae of Florida. I. The intertidal rocks at Marineland. Fla. State Univ. Stud. no. 7 (Papers from the Oceanographic Institute), pp. 17–23.

————. 1962. Marine algae of Virginia as a source of agar and agaroids. Spec. Sci. Report no. 37, Va. Inst. Mar. Sci., Gloucester Point. 13 pp.

————. 1962–63. Agar from Florida seaweeds. Jour. Fla. Assoc. Sci. Teachers. 6(3):11–13; 6(4):11–13, 19, 21.

————. 1963a. *Dictyota dichotoma* in Virginia. Va. Jour. Sci., n. s., 14:109–11.

————. 1963b. Some new records and range extensions of Florida marine algae. Bull. Mar. Sci. 13:516–26.

Humm, H. J. 1964. Epiphytes of the seagrass, *Thalassia testudinum*, in Florida. Bull. Mar. Sci. 14:306–41.

———. 1969. Distribution of marine algae along the Atlantic coast of North America. Phycologia 7:43–53.

———. 1976. The benthic algae of Biscayne Bay. Proc. Biscayne Bay Symposium I. Univ. of Miami Sea Grant Spec. Report no. 5, pp. 71–93. Miami.

———, and R. L. Caylor. 1957. The summer marine flora of Mississippi Sound. Publ. Inst. Mar. Sci., Univ. of Texas 4:228–64, 9 pls.

———, and R. M. Darnell. 1959. A collection of marine algae from the Chandeleur Islands. Publ. Inst. Mar. Sci., Univ. of Texas 6:265–76.

———, and D. Hamm. 1976. New records and range extensions of benthic algae in the Gulf of Mexico. Fla. Sci. 39:42–45.

———, and H. H. Hildebrand. 1962. Marine algae from the Gulf coast of Texas and Mexico. Publ. Inst. Mar. Sci., Univ. of Texas 8:227–68.

———, and C. R. Jackson. 1955. A collection of marine algae from Guantanamo Bay, Cuba. Bull. Mar. Sci. 5:240–46.

———, and Sylvia Earle Taylor. 1961. Marine Chlorophyta of the upper west coast of Florida. Bull. Mar. Sci. 11:321–80, 17 figs.

———, and L. G. Williams. 1948. A study of agar from two Brazilian seaweeds. Amer. Jour. Bot. 35:287–92.

Jao, C.-C. 1936. New Rhodophyceae from Woods Hole. Bull. Torrey Bot. Club 63:237–58.

Jarosch, R. 1962. Gliding. Pp. 573–81, chap. 36, *in* R. A. Lewin (editor), Physiology and Biochemistry of Algae. New York.

Joseph, E. B., W. H. Massman, and J. J. Norcross. 1960. Investigations of inner continental shelf waters off lower Chesapeake Bay. I. General introduction and hydrography. Chesapeake Sci. 1:155–67.

———, ———, and ———. 1961. Hydrographic data from the Atlantic plankton cruises of the R/V *Pathfinder*, December, 1959–December, 1960. Spec. Sci. Report no. 18, Va. Inst. Mar. Sci., Gloucester Point. 2 pp, 12 figs, 2 tables.

Juergens, G. H. B. 1816–22. Algae aquaticae quas in littore maris Dynastium Jeveranam et Frisiam orientalem alluentis rejectas et in harum terrarum aquis habitantes collegit. Decades 1–20. Jever.

Kapraun, D. F. 1970. Field and cultural studies of *Ulva* and *Enteromorpha* in the vicinity of Port Aransas, Texas. Contrib. Mar. Sci., Univ. of Texas 15:205–85.

———. 1977. Studies on the growth and reproduction of *Antithamnion cruciatum* in North Carolina. Norw. Jour. Bot. 24:269–74.

———. 1978. Field and culture studies on growth and reproduction of

Callithamnion byssoides in North Carolina. Jour. Phycol. 14:21–24.

Kemp, A. F. 1960. A classified list of marine algae from the lower St. Lawrence. Canadian Nat. 5:30–42.

Kim, C. S. 1964. Marine algae of Alacran Reef, southern Gulf of Mexico. Ph.D. thesis, Duke University. x plus 213 pp., 7 pls.

————, and H. J. Humm. 1965. The red alga, *Gracilaria foliifera,* with special reference to the cell wall polysaccharides. Bull. Mar. Sci. 15: 1036–50.

Kjellman, F. R. 1872. Bidrag till kännedomen om Skandinaviens Ectocarpeer och Tilopterider. Akad. Afhandling. Upsala. 112 pp., 2 pls.

————. 1877. Über die Algenvegetation der Murmanschen Meeres an der Westküste von Nowaja Semlja und Wajgatsch. Nova Acta Regiae Soc. Sci. Upsaliensis, ser. 3, vol. extraord. no 12. Upsala.

————. 1883. The Algae of the Arctic Sea. Kongl. Svenska Vetenskaps-Akademiens Handlingar, vol. 20, no. 5. Stockholm. 352 pp., 31 pls.

————. 1890. Handbok i Skandinaviens hafsalgflora. I. Fucoideae. Stockholm. iv plus 103 pp., 17 figs.

Koeman, R. P. T., and A. M. Cortel-Breeman. 1976. Observations on the life history of *Elachista fucicola* in culture. Phycologia 15:107–18, 22 figs.

Kuckuck, P. 1891. Bermerkungen über die algenvegetation von Helgoland. Wiss. Meeresuntersuch. N. F. 1(1):223–263.

Kuckuck, P. 1956. Ectocarpaceen-Studien IV. *Herponema, Kützingiella nov. gen., Farlowiella nov. gen.* Herausgegeben von Peter Kornmann. Helgoländer Wiss. Meeresunters. 5:292–325, 15 figs.

Kuntze, O. 1898. Revisio generum plantarum vascularium omnium atque cellularium multarum secundum leges nomenclaturae internationales cum enumeratione plantarum exoticarum in itinere mundi collectarum. Part 3, (section) 2. Wurtzburg. vi plus 202 plus 576 pp.

Kützing, F. T. 1843. Phycologia generalis, oder Anatomie, Physiologie, und Systemkunde der Tange. Leipzig. xxxiii plus 458 pp., 80 pls.

————. 1845. Phycologia germanica d. i. Deutschlands Algen in bündigen Beschreibungen, nebst einer Anleitung zum Untersuchen und Bistimmen dieser Gewachse für Anfänger. Nordhausen. x plus 240 pp.

————. 1845–71. Tabulae phycologicae. Vols. 1–20. Nordhausen.

————. 1849. Species algarum. Leipzig. vi plus 922.

Kylin, H. 1907. Studien über die Algenflora der schwedischen Westküste. Akademische Abhandlung. Uppsala. iv plus 288 pp., 7 pls.

————. 1933. Über die Entwicklungsgeschichte der Phaeophyceen. Lunds Univ. Arsskrift, N. F. Avd. 2, 29(7):1–102, 2 pls.

————. 1934. Zur Kenntnis der Entwickslungsgeschichte einiger Phaeophyceen. Lunds Univ. Arsskrift, N. F. Avd. 2, 30(9):1–189.

Kylin, H. 1947a. Die Phaeophyceen der schwedischen Westküste. Lunds Univ. Arsskrift, N. F. Avd. 2, 43(4):1–99, 61 figs., 18 pls.

————. 1947b. Über die Fortpflanzungsverhaltnisse in der Ordnung Ulvales. Kungl. Fysiogr. Sallsk. Forhandl., vol. 17, no. 17. Lund.

————. 1949. Die Chlorophyceen der schwedischen Westküste. Lunds Univ. Arsskrift, N. F. Avd. 2, 45(4):1–79.

Lagerheim, G. 1886. Note sur le *Mastigocoleus,* nouveau genre des Algues marines de l'ordre des Phycochromacées. Notarisia 1:65–69.

Lamouroux, J. V. F. 1809. Exposition des caracteres du genre Dictyota et tableau des espèces qu'il renforme. Jour. Bot. (Rédigé) 2:38–44.

————. 1812. Sur classification des polypiers coralligenes non entierement pierreux. Nouvelle Bull. des Sci. Soc. Philomatique de Paris 3:181–88.

————. 1813. Essai sur la genres de la famille des thalassiophytes non articulees. Ann. Mus. d'Hist. Nat. (Paris) 20:21–47, 115–39, 267–93, pls. 7–13.

————. 1816. Histoire des polypiers coralligenes flexibles, vulgairement nommes Zoophytes. Caen. lxxxiv plus 560 pp., 19 pls.

————. 1825. *Gelidium.* Dict. Class. Hist. Nat., vol. 7. Paris.

Lang, N. J., and P. Fay. 1971. The heterocysts of bluegreen algae. II. Details of ultrastructure. Proc. Roy. Soc. Bot., ser. B, 178:193–203.

Lee, R. K. S. 1969. A collection of marine algae from Newfoundland. Naturaliste Canadien 95:123–45.

Le Jolis, A. 1856. Examen des espèces confondues sous le nom de *Laminaria digitata* auct., suivi de quelques observations sur le genre *Laminaria.* Nova Acta Academiae Cesareae Leopoldino-Carolinae Naturae Curiosorum, vol. 26, Pars Posterior. Vratislaviae et Bonnae.

————. 1863. Liste des algues marines de Cherbourg. Mem. Soc. Imp. Sci. Nat. Cherbourg 10:1–168, 6 pls.

Lemoine, Mme P. 1917. Corallinaceae, Melobesieae. Pp. 147–82 *in* F. Børgesen, Marine algae of the Danish West Indies. Part III. Rhodophyceae. Dansk Bot. Arkiv 3(1c).

Lewis, I. F. 1910. Periodicity in *Dictyota* at Naples. Bot. Gaz. 50:59–64.

Lightfoot, J. 1777. Flora Scotia, or a systematic arrangement, in the Linnean method, of the native plants of Scotland and the Hebrides. London. 1,149 pp.

Lindstedt, A. 1943. Die Flora der marinen Cyanophyceen der schwedischen Westküste. Lund. 121 pp., 11 pls.

Link, H. F. 1820. Epistola de Algis aquaticis in genera disponendis. *In* C. G. Nees von Esenbeck, Horae physicae Berolinenses collectae ex symbolis virorum. Bonn. Pp. 1–8, pl. 1.

Linnaeus, C. 1753. Species plantarum, exhibentes plantas rite cognitas ad genera relatas, cum differentiis specificis, nominibus trivialibus, synonymis selectis, locis natalibus, secundum systema sexuale digestas. Stockholm. 1200 pp.

―――. 1761. Fauna suecica sistens animalia Sueciae regni : mammalia, aves, amphibia, pisces, insecta, vermes. Distributa per classes & ordines, genera & species, cum differentiis specierum, synonymis auctorum, nominibus incolarum, locis natalium, descriptionibus insectorum. Editis altera, auctior. Stockholmiae, sumtu & literis direct. Laurentii Salvii. 1 pl., 24 figs., 578 pp., 2 pl.

Lüning, K., and M. J. Dring. 1973. The influence of light quality on the development of the brown algae *Petalonia* and *Scytosiphon*. Brit. Phycol. Jour. 8 :333–38.

Lyngbye, H. C. 1819. Tentamen hydrophytologiae Danicae, continens omnia hydrophyta cryptogamia Daniae, Holsatiae, Faeroae, Islandiae, Groenlandiae hucusque cognita systematice disposita, descripta et iconibus illustrata, adjectis simul speciebus Norvegicus. Copenhagen. xxxiii plus 240 pp., 70 pls.

McHugh, J. L. 1959. Maximum-minimum water temperatures, York River, Va., 1952–1958. Spec. Sci. Report no 16, Va. Inst. Mar. Sci. 8 pp.

Mangum, C. P., S. L. Santos, and W. R. Rhodes, Jr. 1968. Distribution and feeding in the onuphid polychaete, *Diopatra cupraea* (Bosc.). Marine Biol. 2 :33–40.

Margulis, L. 1968. Evolutionary criteria in thallophytes: a radical alternative. Science 161 :1020–22.

Martius, C. F. P. von. 1833. Flora brasiliensis. Vol. 1. Stuttgart. iv plus 390 pp.

Mason, L. R. 1953. The crustaceous coralline algae of the Pacific coast of the United States, Canada, and Alaska. Univ. Calif. Publ. Bot. 26(4) :313–90, pls. 27–46.

Mathieson, A. C., C. J. Dawes, and H. J. Humm. 1969. Contributions to the marine algae of Newfoundland. Rhodora 71 :110–59.

Meeuse, B. J. D. 1962. Storage Products. Chap. 18, pp. 289–313, *in* R. A. Lewin (editor), Physiology and Biochemistry of Algae. New York and London.

Meyen, J. 1838. Jahresbericht über die Resultate der Arbeiten im Felde der physiologischen Botanik von dem Jahre 1837. Arch. Naturgesch. 4(2) :1–186.

Montagne, J. F. C. 1840. Seconde centurie de plantes cellulaires exotiques nouvelles. Decades I et II. Ann Sci. Nat., Bot., ser. 2, 13 : 193–207.

―――. 1849. Sixième centurie de plantes cellulaires nouvelles, tant indigènes qu'éxotiques. Decades III a VI. Ann Sci. Nat., Bot., ser. 3, 11:33–66.

———. 1856. Sylloge generum specierumque plantarum cryptogamarum. Paris. xxiv plus 498.

Mueller, Theresa M. 1976. The distribution and seasonality of marine algae of coastal Louisiana and the adjacent offshore continental shelf. Master's thesis, University of South Florida, St. Petersburg. 151 pp., 4 figs., 16 tables.

Müller, O. F. 1782. Icones plantarum. fasc. 15, pp. 1–6. Flora Danica. Copenhagen.

Nägeli, C. 1847. Die neueren Algensysteme und Versuch zur Begründung eines eigenen Systems der Algen und Florideen. Zürich. 275 pp., 10 pls.

———. 1861. Beiträge zur Morphologie und Systematik der Ceramiaceae. Situngsb. Königl. Bayerische Akad. Wiss. München, Jahrg. 1861, 2:297–415, 1 pl.

Newton, Lily. 1931. A Handbook of the British Seaweeds. London. xiii plus 478 pp., 270 figs.

Norcross, J. J., W. H. Massman, and E. B. Joseph. 1962. Data on coastal cruises off Chesapeake Bay. Spec. Sci. Report no. 31, Va. Inst. Mar. Sci., 5 pp., 3 figs, 16 tables.

Ott, F. D. 1973. The marine algae of Virginia and Maryland including the Chesapeake Bay area. Rhodora 75:258–96.

Papenfuss, G. F. 1945. Review of the *Acrochaetium-Rhodochorton* complex of the red algae. Univ. Calif. Publ. Bot., 18(14):299–334.

———. 1947. Further contributions toward an understanding of the *Acrochaetium-Rhodochorton* complex of the red algae. Univ. Calif. Publ. Bot., 18(19):433–47.

———. 1950. Review of the genera of algae described by Stackhouse. Hydrobiologia 2:181:208.

———. 1951. Problems in the classification of marine algae. Svensk Bot. Tidskrift 45:4–11.

———. 1961. Structure and reproduction of *Caloglossa leprieurii*. Phycologia 1:8–31.

———. 1968. A history, catalogue, and bibliography of Red Sea benthic algae. Jour. Bot. 17:1–118.

Pearse, A. S., and L. G. Williams. 1951. Biota of the reefs off the Carolinas. Jour. Elisha Mitchell Sci. Soc. 67:133–61.

Pedersen, P. M. 1974. The life history of *Sorocarpus micromorus* in culture. Brit. Phycol. Jour. 9:57–61.

Perkins, R. D., and I. Tsentas. 1976. Microbial infestation of carbonate substrates planted on the St. Croix shelf, West Indies. Bull. Geol. Soc. Amer. 87:1615–28.

Prescott, G. W. 1951. Algae of the Western Great Lakes Area. Bloomfield Hills, Mich. 946 pp., 136 pls.

Pringsheim, N. 1862. Beiträge zur Morphologie der Meeresalgen. Abhandl. Kgl. Akad. Wissensch. 1862:1–37.

Ragan, M. A. 1976. Physodes and the phenolic compounds of brown algae. Composition and significance of physodes *in vivo.* Botanica Marina 19:145–54.

Ralph, R. D. 1977. The Myxophyceae of the marshes of southern Delaware. Ches. Sci. 18:208–21.

Ravanko, O. 1970. Morphological, developmental, and taxonomic studies in the *Ectocarpus* complex. Nova Hedwigia 20:179–252.

Reinke, J. 1879. Zwei parasitische Algen. Bot. Zeitg. 37:473–78, 1 pl.

––––––. 1889–92. Atlas deutscher Meersalgen. Berlin. 104 pp., 50 pls.

Rhodes, R. G. 1970a. Seasonal occurrence of marine algae on an oyster reef in Burtons Bay, Virginia. Chesapeake Sci. 11:61–63.

––––––. 1970b. Relation of temperature to development of the macrothallus of *Desmotrichum undulatum.* Jour. Phycol. 6:312–14.

––––––. 1972. Studies on the biology of the brown algae on the Atlantic coast of Virginia. I. *Porterinima fluviatile.* Jour. Phycol. 8:117–19.

––––––. 1976. Additions to the brown algal flora of the Atlantic coast of Virginia. Chesapeake Sci. 17:177–81.

––––––, and Mary U. Connell. 1973. Biology of brown algae on the Atlantic coast of Virginia. II. *Petalonia fascia* and *Scytosiphon lomentaria.* Chesapeake Sci. 14:211–15.

Rhyne, C. 1973. Field and experimental studies on the systematics and ecology of *Ulva curvata* and *Ulva rotundata.* UNC—Sea Grant Publ. 73–09. 123 pp.

Rosanoff, S. 1886. Recherches anatomique sur les Melobesiées. Soc. Imp. Sci. Nat. et Math., Cherbourg, Mem. 12:1–112, 7 pls.

Rosenvinge, L. K. 1909. The marine algae of Denmark. Part I. Introduction and Rhodophyceae I (Bangiales and Nemalionales). Dansk Vidensk. Selsk Skr. 7 Raekke, Afd., 7:1–151, 72 figs., 2 pls. 2 maps.

––––––. 1923–24. The marine algae of Denmark. Part III. Rhodophyceae III (Ceramiales). Dansk Vidensk. Selsk. Skr. 7 Raekke, Afd., 7:286–486.

––––––. 1935. On some Danish Phaeophyceae. Kgl. Danske Vidensk. Selskab, Skrifter, naturv. og mathem. Afd., 9. Raekke, 6(3):3–12.

––––––, and S. Lund. 1941. The marine algae of Denmark. Vol. II. Phaeophyceae, pt. 1, Ectocarpaceae and Acinetosporaceae. Kgl. Danske Vidensk. Selskab, Biol. Skr. 1(4):1–79, 38 figs.

––––––, and ––––––. 1947. The marine algae of Denmark. Vol. II. Phaeophyceae, pt. 3, Encoeliaceae through Laminariaceae. Kgl. Danske Vidensk. Selskab, Biol. Skr. 4(5):1–99, 33 figs.

Roth, A. W. 1797–1806. Catalecta botanica quibus plantae novae et minus cognitae describuntur atque illustrantur. Leipzig. Fasc. 1, 1797, viii plus 244 pp., 8 pls.; fasc. 2, 1800, x plus 258 pp., 9 pls.; fasc. 3, 1806, viii plus 350 pp., 12 pls.

————. 1798. Nova plantarum species. Roemer, Arkiv fur die Botanik, vol. 1, no. 3. Leipzig.

Sauvageau, C. 1897. Sur quelques Myrionemacées. Ann. Sci. Nat., Bot., ser. 8, 5:161–288, 29 figs.

————. 1927. Sur les problêmes du Giraudya. Bull. Sta. Biol. d'Arcachon 24:3–74, 18 figs.

Schmitz, F. 1889. Systematische Übersicht der bisher bekannten Gattungen der Florideen. Flora 72:435–56, 1 pl.

Setchell, W. A. 1915. The law of temperature connected with the distribution of marine algae. Ann. Missouri Bot. Gard. 2:287–305.

————, and N. L. Gardner. 1925. The marine algae of the Pacific coast of North America. Part III. Melanophyceae. Univ. Calif. Publ. Bot. 8:387–898, 73 pls.

Silva, P. C. 1950. Generic names of algae proposed for conservation. Hydrobiologia 2:252–80.

————. 1952. A review of nomenclatural conservation in the algae from the point of view of the type method. Univ. Calif. Publ. Bot. 25(4):241–324.

————. 1962. Classification of algae. App. A, pp. 827–37, *in* R. A. Lewin, editor, Physiology and Biochemistry of Algae. New York.

Smith, G. M. 1944. Marine Algae of the Monterey Peninsula. Stanford, Calif. vii plus 622 pp., 98 pls.

————. 1950. Fresh-water Algae of the United States. 2nd ed. New York. vii plus 719 pp., figs.

————. 1955. Cryptogamic Botany. Vol. 1. New York. viii plus 545 pp., 299 figs.

Sprengel, C. 1807. Flora Halens. Mantissa 1:14.

Stanier, R. Y., R. Kunisawa, M. Mandel, and G. Cohen-Bazire. 1971. Purification and properties of unicellular blue-green algae (Order Chroococcales). Bact. Rev. 35:171–205.

Stewart, W. D., P. A. Haystead, and H. W. Pearson. 1969. Nitrogenase activity in heterocysts of blue-green algae. Nature 224 (5216):226–28.

Stockmayer, S. 1890. Über die Algengattung *Rhizoclonium*. Verhandl. k. k. Zool. Bot. Ges., Wien 40:571.

Strickland, J. C. 1940. The Oscillatoriaceae of Virginia. Amer. Jour. Bot. 27:628–33.

Strömfelt, H. F. G. 1884. Om algvegetationen i Finlands Sydvestra Skärgård. Bidrag tell Kännedomen om Finland Natur och Folk, vol. 39. Helsinki.

Taylor, W. R. 1928. The marine algae of Florida, with special reference to the Dry Tortugas. Papers from the Tortugas Lab., Carnegie Inst. of Wash. 25 :1–219, 37 pls.

———. 1957. Marine Algae of the Northeastern Coast of North America. Rev. ed. Ann Arbor, Mich. ix plus 509 pp., 60 pl.

———. 1960. Marine Algae of the Eastern Tropical and Subtropical Coasts of the Americas. Ann Arbor, Mich. ix plus 870 pp., 14 photos, 80 pls.

Thuret, G. 1875. Essai de classification des Nostochinées. Ann. Sci. Nat., Bot., ser. 6 1 :372–82.

Tilden, Josephine E. 1910. Minnesota Algae, vol. 1. The Myxophyceae of North America and Adjacent Regions Including Central America, Greenland, Bermuda, the West Indies, and Hawaii. Minneapolis. iv plus 319 pp., 20 pls.

Turner, D. 1808–19. Fuci sive plantarum fucorum generi a botanicis ascriptarum icones descriptiones et historia. London. 1808, vol. 1, 164 pp., 71 pls.; 1809, vol. 2, 162 pp., 63 pls.; 1811, vol. 3, 148 pp., 62 pls.; 1819, vol. 4, 153, pp., 62 pls.

van den Hoek, C. 1963. Revision of the European Species of *Cladophora*. Leiden. vii plus 248 pp., 55 pls.

Vickers, Anna. 1905. Liste des algues marines de la Barbade. Ann. Sci. Nat., Bot., ser. 9, 1 :45–66.

———, and Mary Helen Shaw. 1908. Phycologia Barbadensis. Privately printed, Paris. 44 pp., 87 pls.

Villalard, M. 1967. Les Ulvacées tubuleuses et filamenteuses de la Baie des Chaleurs et de la Baie de Gaspé (Quebec). Naturaliste Canadien 94 :359–66.

West, G. S., and F. E. Fritsch. 1927. A Treatise on the British Freshwater Algae. Rev. ed. Cambridge. 534 pp.

Whittick, A., and R. G. Hooper. 1977. The reproduction and phenology of *Antithamnion cruciatum* in insular Newfoundland. Canadian Jour. Bot. 55 :520–24.

Wilce, R. T. 1959. Marine algae of the Labrador peninsula and northwest Newfoundland. Nat. Mus. of Canada Bull. no. 158, 103 pp., 11 pls.

Wille, J. N. F. 1880. Bidrag til Kundskaben on Norges Ferskvandsalger, I. Christiania Vidensk. Selsk. Forh., no. 11.

Wille, N. 1901. Studien über Chlorophyceen. I–VII. Vidensk. Selsk. Skr. Christiania, Math.-Nat. Kl., 1900(6) :1–46, 4 pls.

Williams, J. L. 1905. Studies in the Dictyotaceae. III. Periodicity of the sexual cells in *Dictyota dichotoma*. Ann. Bot. 19 :531–60.

Williams, L. G. 1948. Seasonal alternation of marine floras at Cape Lookout, N.C. Amer. Jour. Bot. 35 :682–95.

Willis, W. M. 1973. Phycosociology of the estuarine eulittoral of Smith Island, Virginia: a quantitative approach to seasonal periodicity. Master's thesis, Institute of Oceanography, Old Dominion University, Norfolk, Va.

Wiseman, D. R. 1966. A preliminary survey of the Rhodophyta of South Carolina. Master's thesis, Duke University, Durham, N.C.

Wood, R. D., and E. A. Palmatier. 1954. Macroscopic algae of the coastal ponds of Rhode Island. Amer. Jour. Bot. 41:135–42.

Woodward, T. J. 1794. Description of *Fucus dasyphyllus*. Trans. Linn. Soc. 2:239, 3 figs., 1 pl.

Woronin, M. 1869. Beitrag zur Kenntnis der Vaucherien. Bot. Zeitg. 27(10):columns 153–162, 2 pls.

Wulfen, F. X. 1789. Plantae rariores Carinthiacae, pt. 2. P. 146 *in* N. J. Jacquin, Collectanea ad botanicam, chemiam et historiam naturalem spectantia, vol. 3.

——. 1803. Cryptogamia Aquatica. Leipzig. 64 pp., 1 pl.

Wulff, B. L., and K. L. Webb. 1969. Intertidal zonation of marine algae at Gloucester Point, Virginia. Chesapeake Sci. 10:29–35.

——, Ella May Wulff, B. H. Robinson, J. K. Lowry, and H. J. Humm. 1968. Summer marine algae of the jetty at Ocean City, Maryland. Chesapeake Sci. 9:56–60.

Wynne, M. J., and W. R. Taylor. 1973. The status of *Agardhiella tenera* and *Agardhiella baileyi*. Hydrobiologia 43:93–107.

Zaneveld, J. S. 1965. The benthic marine algae of Virginia. Va. Jour. Sci. 16:346.

——. 1966. The marine algae of the American coast between Cape May, N.J., and Cape Hatteras, N.C. I. The Cyanophyta. Botanica Marina 9:101–28.

——. 1972. The benthic marine algae of Delaware, U.S.A. Chesapeake Sci. 13:120–38.

——, and W. D. Barnes. 1965. Reproductive periodicities of some benthic algae in lower Chesapeake Bay. Chesapeake Sci. 6:17–32.

——, and W. M. Willis. 1974. The marine algae of the American coast between Cape May, N.J., and Cape Hatteras, N.C. II. The Chlorophycophyta. Botanica Marina 17:65–81.

——, and ——. 1976. The marine algae of the American coast between Cape May, N.J., and Cape Hatteras, N.C. III. The Phaeophycophyta. Botanica Marina 19:33–46.

Zollinger, H. 1854. Systematisches Verzeichniss der im indischen Archipel, in den Jahren 1842–1848 gesammelten sowie der aus Japan empfangenen Pflanzen. Zürich. Vol. 1. xii plus 80 pp.

INDEX